Lessons from Deregulation

Lessons from Deregulation

Telecommunications and Airlines after the Crunch

Alfred E. Kahn

AEI-BROOKINGS JOINT CENTER
FOR REGULATORY STUDIES
Washington, D.C.

Lessons from Deregulation may be ordered from:
Brookings Institution Press
1775 Massachusetts Avenue, N.W.
Washington, D.C. 20036
Tel.: 1-800-275-1447 or (202) 797-6258
Fax: (202) 797-6004
www.brookings.edu

Library of Congress Cataloging-in-Publication data

Kahn, Alfred E. (Alfred Edward)
    Lessons from deregulation : telecommunications and airlines after the crunch / Alfred E. Kahn.
        p.   cm.
    Includes bibliographical references and index.
    ISBN 0-8157-4819-1 (pbk. : alk. paper)
    1. Aeronautics, Commercial—Deregulation—United States.
2. Airlines—Deregulation—United States. 3.Telecommunication—
Deregulation—United States. I. Title.
HE9803.A3K28 2004
3849.041—dc22   2003022922

9 8 7 6 5 4 3 2 1
The paper used in this publication meets minimum requirements of the American National Standard for Information Sciences—Permanence of Paper for Printed Library Materials: ANSI Z39.48-1992.

Typeset in Adobe Garamond

Composition by Circle Graphics
Columbia, Maryland

Printed by R. R. Donnelley
Harrisonburg, Virginia

Contents

Foreword

Fred Kahn is one of a kind. As chairman of the New York Public Service Commission, he ignited the quest for more efficient regulation of telephone and electricity monopolies. As chairman of the federal Civil Aeronautics Board, he opened the door to airline deregulation. And as one of the most insightful and articulate proponents of economic deregulation, he helped set the agenda for regulatory reform around the globe.

In this book, Dr. Kahn examines the role of regulation (or lack thereof) in the recent financial meltdowns in the airlines and telecommunications industries. He concludes that the catastrophic outcomes experienced by investors in these industries did not result from deregulation. Indeed, he argues that more deregulation, not less, is the order of the day.

This volume is one in a series commissioned by the AEI-Brookings Joint Center for Regulatory Studies to contribute to the continuing debate over regulation. The series addresses several fundamental issues in regulation, including the design of effective reforms, the impact of proposed reforms on the public, and the political and institutional forces that affect reform. We hope that this series will help illuminate many of the complex issues involved in designing and implementing regulation and regulatory reforms at all levels of government.

The views expressed here are those of the author and should not be attributed to the trustees, officers, or staff members of the American Enterprise Institute or the Brookings Institution.

ROBERT W. HAHN
Executive Director
ROBERT E. LITAN
Director
AEI-Brookings Joint Center for Regulatory Studies

Acknowledgments

I gratefully acknowledge the invaluable comments of Jonathan Baker, Gary Dorman, Michael E. Levine, Dennis Weisman, John C. Wohlstetter, Charles A. Zielinski; the research assistance of Jaime D'Almeida and Peter Rothemund; the continuing support of Richard Rapp and National Economic Research Associates (NERA); the superb editing of Peter Passell; and Martha Ullberg and Tanis Furst for taking charge of the editorial process. The section on telecommunications draws so heavily on my previous collaborations with Timothy J. Tardiff and has profited so from his continuing assistance and review that mere acknowledgement of his contribution is inadequate. Still, coauthorship would unfairly hold him responsible for its deficiencies.

Lessons from Deregulation

1 | *Introduction*

It's no secret that during the last two to three years both the airline and the telecommunications industries have experienced catastrophic declines in the value of their securities.[1] Since these industries were among the most important—and most visible—to have been unleashed from regulation in recent decades (albeit in widely differing degree), their wrenching experience has understandably raised the question of whether their deregulation should be reconsidered or even reversed.[2]

The airlines were comprehensively deregulated in 1978 in one bold stroke. Six years later, the Civil Aeronautics Board (CAB), the government apparatus for controlling domestic fares and routes, was abolished. And although the wisdom of that change is still disputed—not least because of the hard times the airlines have experienced since the economic boom of the 1990s ended—Congress is not about to reverse the process.[3]

Telecommunications is in the midst of a parallel initiative, but one that is both more gradual and more complex. Perhaps most significant, the progression is being comprehensively managed by the regulatory agencies themselves.[4] Indeed, as J. Gregory Sidak points out, in the purported deregulation arena, less often translates as more: the number of pages in the official compendium of Federal Communications Commission (FCC) decisions and proceedings has nearly tripled since passage of the Telecommunications Act of 1996, while membership in the Federal Communications Bar Association increased by 73 percent between December 1994 and December 1998 and has remained essentially at that level.[5]

There are, of course, reasons, good and bad, for the sharp differences in the course of deregulation and deregulatory policies in these two industries—reasons of history, technology, and politics. What there is in common is the successful demonstration of the superiority of open competition over direct comprehensive regulation. In my view, however, every passing year demonstrates also the superiority of the road we chose for the airlines and—I think it not an exaggeration to say—the bankruptcy of the highly managed or regulated course we have taken in telecommunications.

The financial collapse of these two industries did not, of course, take place in isolation. Technology-related stocks in general, and the dot-coms in particular, suffered at least as dramatic a meltdown.[6] Since these latter companies have essentially been free of direct economic regulation throughout, their experience provides a useful counterpoise to the natural tendency to blame all the woes of aviation and telecommunications on government policy. (There is obviously a separate story to be told of derelictions, whether of the government or private groups, in the enforcement of non-industry-specific prescriptions of accounting and financial reporting standards and in the prohibition of simple fraud, which apparently played a very important role in the cases of both dot-coms and telecoms.)[7]

2 | *The Airlines: "Normal" Recession plus 9/11 and Iraq*

I begin by baldly stating my essential conviction: airline deregulation has been a nearly unqualified success, despite the industry's unusual vulnerability to recessions, acts of terrorism, and war. The carriers' freedom to enter (and exit) domestic markets and to price as they please, subject only to the constraints of the antitrust laws, has generated enormous benefits for consumers. The most comprehensive study of the experience, by Steven Morrison of Northeastern University and Clifford Winston of the Brookings Institution, estimates that these benefits exceed $20 billion annually.[8] These take the form primarily of lower fares and vast increases in the availability of one-stop service between hundreds of cities—the latter made possible by the carriers' new freedom to operate hub-and-spoke route systems.

True, lower fares have come at the cost of deterioration in the quality of the typical customer experience. In the decade before deregulation, domestic flights were, on average, less than 53 percent full; in 1997–2001, they averaged over 70 percent. But crowding reflects the success of deregulation, not its failure. Competition in the unregulated market has proved to the satisfaction of the carriers that most travelers are willing to sacrifice comfort for lower fares.[9]

Cyclical Sensitivity plus 9/11

The demand for air travel has always been unusually sensitive to changes in general business conditions, and the heavy fixed costs of offering convenient flight schedules have discouraged prompt corresponding curtailment of supply during downturns. Hence, it should not be surprising that deregulation has led to dramatic price cutting and severe financial distress during general business recessions.[10] The U.S. industry lost some $13 billion in the mild economic doldrums of the early 1990s, exacerbated by war and soaring fuel prices—more, the airlines claimed, than the total profits of the industry since the Wright Brothers' first flight.

In retrospect, it was clear that the major carriers had reacted to the very sharp increases in traffic in the prosperous mid-1980s by ordering more new equipment than they needed. Once the rate of increase of traffic diminished in the early 1990s, they did what players in most competitive industries do when faced with heavy fixed costs and lagging demand: they dropped prices below long-run average costs. The economists' response to the cries for government intervention—namely, that this was a normal cyclical phenomenon and that the industry would learn from previous mistakes—was clearly vindicated by the sharp recovery of the late 1990s. Indeed, from 1995 through 1999 the industry was satisfactorily profitable: its return on investment averaged 12.4 percent.

Set aside for the moment the enormous costs of the security precautions imposed on the industry after 9/11—costs that arguably should be borne by government. One could still believe that the cumulative impact of disasters ranging from terrorism to SARS justifies temporary financial assistance to sustain the nation's air transport infrastructure. By the same token, it seems entirely proper for the offer to be made contingent on major give-backs of the extraordinarily inflated wage costs achieved by the industry's powerful unions—give-backs precisely analogous to the concessions demanded by Congress in exchange for the Chrysler loan guarantee during the Carter administration.

But the unusual vulnerability of an industry to external shocks does not constitute a legitimate case for a return to regulated cartelization. Indeed, the one major surviving restraint on competition, the prohibition against foreigners owning more than 25 percent of domestic carriers, is especially anomalous when the U.S. industry so badly needs additional financing and foreign airlines and entrepreneurs are apparently eager to invest.[11] Ironically, the United States, which has been prodding other countries to "exchange liberalizations for liberalizations" in the form of freer access to the U.S. market for *international* travel ever since the late 1970s, may now be subject to similar pressures from Europe. Members of the European Union, bargaining as a single entity, are now pressing for mutual liberalization in the form of reciprocal rights to compete for *domestic* traffic.[12]

The Triumph and Vulnerability of Hub-and-Spoke Operations

Nor has the case for deregulation been undermined by the other apparent evolutionary force pushing the major carriers to the brink: the weakening of the competitive advantages associated with hub-and-spoke route structures. Hub-and-spoke operations, which make it possible for carriers to offer travelers from small cities one-stop access to hundreds of destinations worldwide and travelers in larger cities much more convenient scheduling than would otherwise be possible, entail huge fixed costs.[13] The resulting economies have never translated into sustained high profitability, because of the exposure of such complex network operations to wage demands by strong unions and because of their increasing subjection in recent years to competition from low-cost, point-to-point carriers on heavily trafficked routes.

Soon after deregulation it became clear that the high-cost network carriers could survive only by employing sophisticated yield management techniques—techniques to extract the maximum revenue per seat by charging different rates to different classes of passengers, according to their demand elasticities. As low-fare carriers like Southwest increasingly siphoned off price-sensitive customers, and the requisite differential between high- and low-elasticity travelers therefore widened, the high-cost hub-and-spoke airlines finally encountered growing resistance from business travelers, increasingly resentful of the seemingly outrageous full fares they were being asked to pay.[14] Many were galvanized to drive long distances to non-hub airports in order to take advantage of the low prices offered by point-to-point carriers. And this trend, accelerated by economic recession, has clearly tipped the balance of advantage away from very high fixed cost hub-and-spoke operators: low-fare carriers reportedly carry a quarter of the passengers today, as contrasted with less than 10 percent a decade ago.[15]

But just as proof of the superiority of hub-and-spoke operations required a test in a market unfettered by regulation, the possibility that hub-and-spoke systems have passed their peak must be tested through open competition with point-to-point operations. The fact that the evolution of airline route and rate structures has been disrupted by recession, terrorism, and war is in no sense an argument for returning control to some inherently incompetent government authority. Just as the CAB effectively obstructed the industry's realization of economies from hub-and-spoke operations—once deregulated, the industry quickly moved to the new model—so, today, it is inconceivable that any governmental authority would be capable of redesigning the industry to comport with the ever-changing realities of the market.[16]

The Restructuring Response

The more relevant question today is to what extent the airlines should be left to seek salvation individually (with or without the protection of bankruptcy), to what extent collectively (with or without the benefit of exceptions from the antitrust laws). In fact, the extreme financial hardship imposed on the industry by recession, the need for intensified security, and the changing balance between point-to-point and hub-and-spoke route structures have understandably led the major incumbents to seek salvation or strength through merger and code sharing.[17] And this inevitably raises the question whether such efforts violate the letter or spirit of the antitrust laws.

Initially, the industry requested exemption from those laws to permit its members to act collectively to reduce their operations in response to the combination of recession and September 11. Since that would have entailed a complete reversal of deregulation—one of the precipitating causes of which was a similar collective response to the 1974–75 recession—those requests were, properly, abandoned.

Even code sharing—agreements to co-brand flights, particularly on overlapping routes—seems inherently anticompetitive, at least potentially, since it is an alternative to direct horizontal rivalry among alliance partners. This is particularly true on routes originating at the hub of one of the alliance members and terminating at the other's, but also holds on end-to-end routes where competition between them might otherwise emerge. There seems every reason to ask whether alliances also exacerbate the competitive disadvantages of incumbent nonmembers (and raise obstacles to independent entry), by increasing dominance of the members at their respective hubs: alliance partners, as I understand, have generally refused to interline (accept tickets from and exchange baggage) with outsiders.[18]

This was presumably the basis for the opposition of Delta, Continental, and Northwest Airlines, as well as Southwest, to the proposed merger of United Airlines (UAL) and US Airways. The question of what weight should be placed on the opposition of would-be competitors to such combinations is, of course, a familiar one in the world of antitrust. Skeptics are inclined to interpret such opposition as further proof of the promised efficiencies of such mergers: of course nonparticipants would object to the intensified (and socially beneficial) competition to which these combinations would expose them. But Delta, Continental, and Northwest objected to the UAL–US Airways merger, persuasively, on the ground that it would give the merging parties strategic advantages unrelated to their efficiency, such as preferential access to traffic originating with the other partner.[19] So an evaluation of their own recently approved alliance must appraise similar objections of their competitors.[20]

Arguing powerfully on the other side, however, has been the dramatic reversal of the industry's fortunes and the apparent inevitability of substantial curtailment in either the number of hub-and-spoke carriers or the scope of their separate operations. I had argued, in opposition to the UAL-US Airways merger, that if United felt its ability to compete in the Northeast would be greatly improved by the acquisition of US Airways' Pittsburgh hub, the more competitive course would be to construct a hub of its own.[21] But that was before September 11 and the Iraq War, which cast my earlier suggestion as ancient history.

Alliances

Since the Department of Justice's Antitrust Division blocked the United-US Airways merger, the two entered an alliance (and each subsequently declared bankruptcy). The government then approved an alliance between the relatively—but only relatively—healthy Continental, Delta, and Northwest, the third, fourth, and fifth largest carriers, respectively. These events raise a delicate, but crucial, question: might such alliances be the least anticompetitive way of effecting the shrinkage of hub-and-spoke operations, needed because these capital-intensive systems have lost their competitive edge? Specifically, might they be less anticompetitive than simple withdrawal from particular routes by one or another of the partners, since staying in the market in an alliance would preserve more of the continuing economies of scope—the ability of each carrier to reach a larger number of geographically far-flung destinations—of competing hub-and-spoke operations?

This is not to assert that the purpose of the recent Continental-Delta-Northwest alliance is indeed to reduce the scope or scale of the combined operations of these carriers. There is no reason to doubt that its purpose is to improve their combined value by (a) achieving or preserving greater economies of scope than they could do individually; (b) to improve the "seamlessness" (the industry's equivalent of cleanliness in its propinquity to Godliness) of their multiple offerings, by rearranging their airport facilities;[22] and (c) increasing their effectiveness by combining frequent flyer programs and airport clubs. It is to suggest, also, that in the altered financial condition of the industry in general and the hub carriers in particular, damping a negative trend is equivalent to enhancing a positive one: simple code-sharing agreements have the virtue of allowing carriers to offer the benefits of hubbing—the much wider range of origins and destinations on a single ticket, at a single combined fare—more widely than would otherwise be possible.[23]

Code sharing also has the virtue, predicted by economic theory and validated by experience, of producing lower fares from origin to destination. That is because carriers that set fares independently for each segment have no incentive to factor into their calculations the potential gains to other airlines associated with feeding traffic into the others' systems.[24] Studies by Brueckner and Whalen conclude that such "pricing synergies" generated by existing airline alliances have produced 18 to 20 percent reductions in fares on interline routes.[25]

Is there reason to be more permissive of proposed alliances between hub-and-spoke carriers today than, say, five years ago? I believe so. During the period that witnessed the triumph of hub and spoke, the public was especially concerned about preserving the opportunity for low-cost point-to-point carriers to challenge the new system—particularly in light of the mounting evidence of the big system carriers' dominance at hub airports and the consequent need for a competitive mechanism to assure that nondiscretionary travelers would be charged no more than the stand-alone costs of serving them.[26]

With the current shrinkage of hub operations, it seems to me equivalently important to permit the hub-and-spoke carriers to preserve and enhance the economies of scope inherent in their particular mode of operations. These economies have lowered prices for customers with highly elastic demand (think, vacationers) and improved the frequency of service for customers with low price elasticities of demand (think, business travel). While we must continue to be concerned about the airlines' illicit accumulation and exercise of exclusionary power, it is not in the interest of the public to sacrifice those efficiencies. And while we must not allow the desperate financial plight of the major carriers to be used as a justification for cartelization, the notion that every existing major carrier must retain complete independence seems unrealistic.

Does this mean that I want antitrust officials to handicap the competitive contest in the way the Civil Aeronautics Board did, permitting companies greater or lesser leeway to suppress competition, depending upon their economic health? I plead not guilty: my mentor, Myron W. Watkins, brought me up to believe that the *Appalachian Coals* decision of 1933, which allowed coal producers to organize into a marketing cartel, was a bad one—grounded in the same syndicalism as was embodied in the National Recovery Administration, as well as the even more explicit Guffey Coal, Motor Carrier, and Civil Aeronautics Acts of 1954, 1935, and 1938. In assessing proposed joint ventures such as the Continental-Delta-Northwest alliance, however—collaborations that assertedly do not directly limit competition among the parties, but confer advantages stemming only from the improved services they can collectively offer—I see no reason to condemn them per se. At the same time, I think antitrust considerations counsel care from the very outset, to eliminate any advantages to the partners—such as combining frequent flyer rebate programs—that are unrelated to efficiencies or improvements in the inherent quality of their services.

The foregoing ruminations raise a big question about the Continental-Delta-Northwest alliance. To what extent will it suppress competition among the partners (by collusion), or with nonmembers by handicapping their ability to compete on the basis of their own efficiencies?[27]

As to the former danger, Michael E. Levine argued that the alliance was indeed fashioned to expand economics of scope without suppressing competition:

Alliances can involve coordination of pricing and output, in which case they present competition issues that should be evaluated in merger terms. But alliances can operate without such coordination. . . . As the proposal is structured there is no common bottom line and no revenue pooling, so no airline can earn passenger revenue unless it carries the passenger itself. This feature alone means that there is no legal way to suppress competition among the partners, since only illegal side payments or reciprocal agreements could compensate them for traffic they "concede" to their partners.

This alliance, he contended, did not "involve coordination of pricing and output, in which case [it would] present competition issues that should be evaluated in merger terms. But alliances can operate without such coordination, offering additional scope [that is, benefits of the economies of scope made possible by hub-and-spoke operations] to a public that prefers it while leaving the alliance partners as competitors."[28]

In approving the Continental-Delta-Northwest alliance, the Justice Department's Antitrust Division agreed about its promised benefits: "[It] has the potential to lower fares and improve service for passengers in many markets throughout the country. . . . Corporations can also benefit from joint bids for contracts from alliance airlines where the airline parties offer complementary rather than competing service." It formally attached conditions, already accepted by the applicants—conditions such as those Michael Levine described with approval—that strengthen the assurances against clear violations of the antitrust laws: "One condition prohibits the carriers from code-sharing on each other's flights wherever they offer competing nonstop service, such as service between their hubs. The conditions also require the carriers to continue to act independently when setting award levels or other benefits of their own frequent flier programs and when they are competitors for corporate contracts."[29]

These assertions provide a useful framework for raising questions about which I confess I have been able to satisfy myself only in part. Do not the alliance code-sharing arrangements, which allow alliance members to put passengers on multisegment flights that originate on the planes of their alliance partners, necessarily imply an agreement to curtail their own overlapping operations? How can such passenger allocation be accomplished except by some sort of agreement about which flights each member will retain, carrying passengers originating with the other, and on which routes they will curtail their own operations and book customers on partners' flights? Or if, as defenders of the Continental-Delta-Northwest alliance claim, it entails no explicit scheduling curtailments, is this still not an inherently likely consequence? And why would it not constitute a division of markets, per se illegal under the antitrust laws?

If, however, as the Justice Department's press release states, "there will be no sharing or pooling of revenues, so each carrier will continue to compete for passengers," it is not clear how one of the major aforementioned advantages of code sharing—combined fares for multiple-segment flights lower than the sum of the fares set individually—can be realized.[30] According to Edward Faberman, executive director of the Air Carriers Association of America, the alliance will involve, "Coordinated inventory management—The three carriers will cooperate in a system to manage and sell each other's inventory so as to maximize total sales."[31] It is not clear how such an agreement would be compatible with the foregoing promise.

The same is true of the inclusion, among Faberman's list of the alliance's provisions, of "Coordinated sales and marketing—the carriers intend to coordinate their sales and marketing programs; presumably this means they will aggregate their travel agency commission override and corporate incentive programs."[32] Can fellow-members of such a comprehensive partnership be expected to price as independently on competitive routes as if they were not allied in this way?[33]

On the other hand, recognizing at the outset that the conditions affirmed by the Department of Justice limit such practices in situations where alliance members offer directly competitive nonstop service, alliances are by their nature intended to offer packages of complementary services—for example, single-ticketed multiple-segment routes and corporate discount programs—more attractive than what the partners could offer individually. The fact that by doing so the partners make it correspondingly more difficult for non-members to compete surely cannot constitute a basis for their condemnation as an unfair method of competition.

Should that exemption apply, however, to mergers of frequent flier programs and airport clubs? Frequent flyer credits were a brilliant competitive innovation. Since the benefits are typically nonlinear—they increase disproportionately with the total accumulated mileage—they offer strong inducements for travelers to concentrate their flights on the airline that offers the greatest number and variety of flights where they live. And this contributes powerfully to hub dominance. Entirely apart from the genuine improvements in service made possible by alliances, rewards for exclusive patronage make it difficult for competitors that are in all other respects equally efficient to challenge them—not only on routes to and from hubs, but in spoke cities as well.[34] The question, then, is whether the combining of those benefits with alliances—while doubtless attractive and beneficial to travelers—would not also create or reinforce monopoly power.[35] I use the term in the historical antitrust sense of power not only over price but to exclude otherwise equally efficient competitors.[36]

"Regulatory" versus "Antitrust" Protective Conditions

I have elsewhere expressed approval of the increasingly common practice on the part of our antitrust enforcement agencies of conditioning merger approvals on the applicants' acceptance of often-complex *structural* changes in their project:

> Despite its substitution of negotiation for litigation, that practice seems
> consistent with the spirit of the antitrust laws: Section 7 of the Clayton Act
> concentrates on the possibly anticompetitive effects of changes in an indus-
> try's structure consequent on mergers: what the antitrust agencies attempt
> to do in those quasi-regulatory negotiations is, in principle, identical—to
> eliminate the aspects of the restructuring effected by the merger that in their
> judgment threaten competition.[37]

Setting aside the questions of their *adequacy*, on the one hand, and their likely thwarting of the full benefits of vertically integrated pricing, on the other, the conditions attached by the Department of Justice to its approval of the Continental-Delta-Northwest alliance seem similarly justified.

In contrast, the Department of Transportation initially condi-tioned its approval of the Continental-Delta-Northwest alliance on the carriers' acceptance of a more intrusive, and in some respects qual-itatively different, set of conditions—illustrating, as I have put it, "the comparative propensities of the antitrust agencies and a regulatory agency to meddle."[38] No doubt this reflected the department's much more skeptical view of the likely benefits from the alliance and, espe-cially relevant in the present context, its greater sensitivity to the likely disadvantages to competitors[39]—a sensitivity arguably dictated by its statutory mandate to prevent unfair methods of competition.[40]

Some of the Department of Transportation's conditions seem fully in keeping with conditions attached by the Department of Jus-tice to merger proposals with the aim of ensuring equality of oppor-tunity for equally efficient competitors:

—the carriers must surrender specified numbers of underused gates at congested airports;

—"the carriers must request that their services be listed under no more than two codes in computer reservation systems (CRS) until the Department completes its pending revision of the CRS rules."[41]

The first is justified by the finding by many investigators, including the General Accounting Office and the department itself, that competitive entry into hub airports, particularly by low-fare carriers, has been hampered by an unreasonable unavailability of gates, and the second by the findings, again by both agencies, that multiple listing of such flights, particularly on the first computer display, has deprived rival airlines of an equal opportunity to compete.[42] In short, they seem to me remedial, in the sense of offsetting or mitigating two exclusionary practices that have impeded competition.

Two others, in contrast, seemed unacceptably prescriptive, aimed not at safeguarding against competition-suppressing behavior by the partners or structural consequences of their alliance but at restricting or dictating future performance:

> *Code-sharing limitations.* In an effort to ensure that the Alliance Carriers fulfilled their promises of consumer benefits due to new on-line service in many markets, we required that at least one-fourth of each marketing carrier's code-sharing flights must be to or from airports that the airline and its regional affiliates either did not directly serve or served with no more than three daily roundtrips as of August 2002. We also required that an additional thirty-five percent of the code-share flights must either meet that requirement or be to or from small hub and non-hub airports. The condition limited the total number of code-sharing flights between Delta and Continental and between Delta and Northwest to 2,600 (but does not affect the existing code-sharing between Continental and Northwest). We committed ourselves to reviewing these restrictions after the first year. We believed these restrictions were necessary to ensure that the Alliance Carriers implemented their representations that the alliance would provide consumer benefits by creating on-line service in a number of new markets. 68 FR 3298

The alternative language allows the Alliance Carriers to code-share on an additional 2,600 flights in the second year, subject to the requirement that thirty percent of these additional code-share flights must be flights in new markets or to small hub or non-hub airports. If the Alliance Carriers wish to add additional code-share flights after the second year, they must give us 180 days advance notice and provide any information requested by us on the additional code-share services. 68 FR 10771

We believe that the alternative language will continue to ensure that the Alliance Carriers use their code-sharing to extend their networks, as they publicly stated was their intent.[43]

Code-sharing, subject to antitrust conditions, either is or is not compatible with healthy competition. If it potentially leads to more effective performance, the kind of limit imposed by the Department of Transportation seems inherently anticompetitive. It is, in my opinion, also incompatible with the principles of deregulation to try to serve social or political ends either by imposing limitations on the markets in which the cooperative arrangements may be used or by obliging the partners to offer particular, specified services. In this instance, the obligation is to increase flights to "unserved or underserved communities" that it would presumably not otherwise be in their interests to serve. This kind of attempt to use the agencies' leverage to force alliance applicants to serve the "public interest" in ways that the regulators would have no authority to require directly seems to me an unacceptable restoration of discredited regulation.[44]

What about Restructuring by Merger?

No doubt because of the Antitrust Division's opposition to the proposed merger of United and US Airways and financial ties between Continental and Northwest, those carriers reframed their proposed collaborations in the form of code-sharing alliances. They also felt constrained to defend the agreements against possible antitrust attack by attaching conditions in terms that make them less likely to be treated either as outright mergers or as pricing or market-allocating agreements between competitors. My foregoing appraisal of the alliances adopted the same frame of reference.

As a lifelong skeptic of the proclaimed cost-saving benefits of most mergers (and one old enough to remember how the union of the Pennsylvania and New York Central Railroads was heralded as their joint salvation), this framework for appraising the alliance makes sense to me. The fact remains, however, that the antitrust laws do not prohibit mergers outright—nor should they.

Indeed, it might appear curious that the financial plight of the major carriers during the last three years, accompanied by the sharply increasing market shares of their lower-cost competitors and the prospect that these trends may indeed reflect a long-term shift in the balance of viability among their respective modes of operation, has not led to renewed initiatives for outright mergers. I attempt no assessment of that hypothetical issue here—an issue made all the more hypothetical by the fact that past mergers have raised monumental problems of integrating work forces with different unions, pay scales, seniority lists, and job descriptions.[45]

In the earlier debates over whether the successes of deregulation were threatened by growing hub dominance, defenders of the emerging network system pointed to the increasing competition among major carriers in long-distance carriage over their respective hubs. Julius Maldutis pointed out, for example, that travelers between Boston and Phoenix at one time had the choice of nine hubs through which to make one-stop connections.[46] According to Orbitz.com, that number stands at nineteen today, with offerings by all seven domestic network carriers. In light of the abysmal financial condition

of the latter companies and the likely long-term shift in the competitive balance toward point-to-point flights—a trend most pertinently
exemplified by the operations that Southwest has since established at
Providence and Manchester, New Hampshire—the apparently uni-
versal refusal to consider outright mergers among the majors seems
anomalous.[47]

3 | *Telecommunications: Tangled Wires and Deregulatory Remedies*

The case of telecommunications is so much more complicated than that of the airlines, I am deterred to the point of muteness—well, not quite—by the challenge of sorting out what ought to happen next.

The Unequivocal Successes of Deregulation

There are some important areas in which the deregulation (or "unregulation") of telecommunications has been extremely successful, although in most cases lamentably delayed.[48]

—Long-distance rates, the overpricing of which bred huge social welfare losses under regulation, are down sharply.[49] A very large share of that price decline was driven not by direct competition among the long-distance carriers but by reductions in the charges regulators required them to pay the incumbent local exchange companies (ILECs) for access to their networks—regulation motivated by the political imperative to subsidize basic residential charges.

Competition did matter a great deal, however: the reductions in local access fees were in large measure forced by the widespread entry of competitive access providers. More rational pricing of long distance must therefore be counted as a major success of deregulation.

—The prices of cellular and other wireless services have declined dramatically, and use is up correspondingly, thanks in large part to unfettered competition.[50]

—Terminal attachments at the subscriber's end (telephone handsets, answering machines, switchboards)—effectively, privately owned local telephone systems—have proliferated. Most important, perhaps, e-mail through computers has provided direct competition with voice telephone service, while the growing use of the Internet for long-distance voice calling now threatens some of the $16 billion a year that the incumbent local companies continue to collect as access fees from the long-distance carriers.[51]

These experiences support several, in part mutually reinforcing, in part mutually contradictory, conclusions:

—Technological change, exemplified in telecommunications, has proved to be the most powerful enemy of regulated monopoly and rate structures distorted by regulation—an incompatibility that is both mutual and reciprocal. This is so even though the ability of a comprehensively regulated monopolist, AT&T, to, in effect, tax ratepayers to support pure as well as applied research led to remarkable scientific progress and technological innovation.

—Regulation, with its corollary goal of preserving those distorted, cross-subsidizing rate structures, has also been the principal obstacle to the rapid exploitation of new technologies. Conversely, the substantial reduction of that cross-subsidization, previously financed by egregious overcharging of long-distance and business services, has been a major source of consumer benefit.[52]

—The elimination of regulatory barriers to competitive entry, even as regulators and incumbents alike attempt to preserve the artificially elevated prices needed to finance cross-subsidies—an attempt promptly abandoned in the case of the airlines[53]—has artificially stimulated competitive entry and is therefore partly responsible for the cycle of boom and collapse over the last several years.

—Again as in the case of airlines, the unintended adverse consequences of deregulation in telecommunications not only offer no good reason to re-regulate and re-cartelize the industry; *they counsel an early abandonment of oxymoronic efforts to promote competition by regulation.*

The Role of the AT&T Divestiture and Its Gradual Reversal

Whether or not the breakup of AT&T and the associated exclusion of the surviving Baby Bells from long-distance service was necessary to generate the benefits of regulatory reform, there can be no dispute that equal access to the facilities of the local companies contributed substantially to the success of competition in long-distance service over the last two decades.[54] There is no need to rehash the arguments over the adequacy of the competition that arose as the local Bells sought the right to offer interLATA (local access and transport area) toll service—notably, the question of whether the declines in long-distance rates were attributable to declines in access charges rather than increases in competition. There can be little question, however, that

—competition did become increasingly effective over time, particularly in serving large business customers;

—the much smaller benefits for small users were largely explained by the higher unit costs of serving them associated with the need to recover the fixed costs of retailing over lower volumes of sales.

By contrast, the readmission of the local Bells into the interLATA arena was particularly beneficial to small customers because the ILECs were in a position to exploit the economies of scope arising from the fact that they were already providing local service to virtually everyone.[55] The long-distance carriers, confronting the much larger incremental costs of marketing and billing associated with taking on small customers, have found it necessary to impose flat monthly charges or comparatively high charges per call.

The experience of Southern New England Telephone (SNET) in Connecticut provides a clear example. SNET began offering out-of-state service in April 1994, at rates 15 and 25 percent below AT&T's undiscounted rates for peak and off-peak calling, respectively.[56] The success of the initiative in bringing the benefits of competition to smaller users is ironically reflected in the testimony of Frederick Warren-Boulton on behalf of AT&T before the Kansas Corporation Commission: in an attempt to minimize the benefit of ILEC reentry, he noted that, while SNET had gained 34 percent of all long-distance customers, it had captured only 12 percent of the revenues.[57]

This experience with the benefits of vertical reintegration—including the internalization of the benefits of expanded sales of complementary services, parallel to those achieved by airline alliances[58]—makes one skeptical about the wisdom of the FCC's systematic insistence in recent years that ventures by the ILECs outside the traditional boundaries of voice service—notably in broadband[59]—be confined to subsidiaries operating at arm's length. The twenty-year experience with AT&T's dissolution should have increased our respect for the potentially large economies of scope in telecommunications.

The Growth of Local Competition

Meanwhile, both state and federal regulators sought to preserve the multibillion dollar per year cross-subsidy between long-distance and basic local service, requiring the local companies to charge the long-distance carriers outrageously inflated prices for access to their local networks.[60] What happened next provided a powerful demonstration of the tendency of technological advance—in the present instance, the development of high-capacity fiber-optic transmission—to undermine rate structures distorted by regulators. By the early 1990s, every major metropolitan statistical area (MSA) in the country had its own competitive access provider, offering both long-distance carriers and large retail telecom customers a means of bypassing the grossly overpriced local facilities of the incumbents: the largest 150 MSAs could boast a total of nearly 1,800 competitive access provider networks.[61]

The Telecommunications Act of 1996 established local competition as a goal in itself—a logical complement to the lifting of the ban on the local Bells offering interLATA services. These competing access providers (CAPs)—the two biggest ones were subsequently acquired by AT&T and MCI WorldCom—formed the backbone of the burgeoning competitive local exchange companies (CLECs), offering a full range of services. Indeed, the top twenty-five MSAs were served by an average of thirty-two CLEC networks *each*. And once again, the dramatic increase in competitive services to business customers was driven by the self-defeating efforts of regulators to subsidize basic services. In fact the CLECs achieved a large double-digit share of that market by investing tens of billions of dollars annually in the latter half of the 1990s.[62] Significantly, of the estimated 16 million to 23 million lines they serve using their own switches, only 3 million or so are residential.[63]

To show that they had complied with section 271 of the 1996 act, opening their local markets to competition sufficiently to justify their own entry into the interLATA market, the local Bell companies (and witnesses for them, such as I) pointed to these huge investments and to the rapidly growing market share of the competitive local service providers. We were well aware that these dramatic expenditures had been artificially stimulated by the grossly distorted rate structures: indeed, we pointed out explicitly that for this reason they did not contradict our contentions that the ridiculously low total element long-run incremental cost (TELRIC) charges prescribed for unbundled network elements (UNEs) were nevertheless discouraging facilities-based local competition (see my digression on the TELRIC pricing issue, below). Why would CLECs be expected to take the risks of constructing their own facilities, we asked rhetorically, when the FCC required the ILECs to make their own network elements available to them at the hypothetical lowest prices that could be achieved by a most efficient competitor, building new facilities from the ground up?[64] Competitive local telephone companies serve approximately 30 percent of all business lines today, but only about 9 percent of residential lines. And of that 9 percent, almost two-thirds are served exclusively with ILEC facilities.[65]

The Bubble Bursts

To what extent can deregulation properly be blamed for the spectacular boom and bust in the telecommunications industry? The massive overinvestments in transport capacity—specifically, long-distance transport capacity—were clearly linked to the introduction of competition in that part of the telecom business. But does that mean deregulation was a failure? On the contrary: as I have already observed, one of deregulation's crowning successes has been the dramatic decline in long-distance rates and the subsequent revolution in customer habits. Dazzling technological progress obviously triggered and inflated both.

A boom and bust in the exploitation of new products and technologies—the combination of technological creativity and speculative excess—is inherent in market capitalism (which is not to say that socialist economies avoid them). Consider the extremity of the cycle driven by the development of "dense wave division multiplexing" (DWDM), which, according to Gregory Sidak, has increased the number of channels attainable from a single strand of fiber from two to over 100, with the possibility it will eventually exceed 1,000.[66] The catastrophic experiences of the essentially unregulated dot-coms and airlines provide a useful perspective: however unhealthy, extreme cycles are an integral aspect of the normal functioning of unregulated markets. They in no way recommend *economic* re-regulation[67]—as distinguished from safeguards against outright fraud and double-dealing by corporate executives and dishonest accounting, auditing, and appraisals by investment analysts and advisers.[68] On the contrary, to the extent that the FCC-imposed sharing obligations discouraged ILEC investments in adding fiber to the local loop (see discussion at notes 92–93, below), it frustrated growth in use of the capacity liberated by the development of DWDM.

As I have already made clear, a very large share of the blame for the boom and subsequent collapse is properly also placed at the door of the grossly distorted rate structures that were the legacy of earlier regulatory policies, and therefore constituted one of the best reasons for deregulation. This criticism is based on a lot more than hindsight.[69]

The result was exuberant overinvestment in both long-distance transport and local distribution facilities, exacerbated by the above-mentioned astronomical progress in fiber transport technology. But the losses have been borne by the investors and their creditors.[70] And properly so, except to the extent to which they were deceived by company executives and auditors; in those cases, market failure was the consequence of lamentably lax enforcement of securities disclosure regulations—not telecommunications deregulation. Booms and busts at the expense of investors seem to be unavoidable costs of a dynamic competitive economy experiencing—and benefiting hugely from—rapid technological progress. The excess physical capacity, once constructed, will not go away, but will be available when demand is sufficient to recover its very low incremental cost. I know of no serious argument that we would be better off today if that part of the business remained a tightly regulated monopoly.[71]

The more immediate (and more fundamental) question is what blame should fall on intrusive *increases* in regulation since 1996, set off by the telecommunications reform act and rationalized as necessary to introduce competition at the local level.

The Apportionment of Blame: The 1996 Telecommunications Act

The climax of six years of intense controversy over the administration of the FCC's responsibilities under the 1996 act—a process that has included rebuffs by both the Supreme Court (in 1999) and the D.C. Circuit Court of Appeals (in 2002), as well as approvals by the former in both years—has understandably set off a torrent of commentary. One side supports the commission's previous actions—indeed, it urges their intensification. The other side bitterly criticizes the FCC's penchant for aggressive promotion of competitors and calls for drastic curtailment.

The FCC's adoption in February 2003 of new rules defining the ILECs' obligations to unbundle network services, which was delayed until the last possible moment by very public internal differences of opinion, offers closure on some issues. But it leaves fundamental questions unresolved, transferring the most difficult ones back to the states, where their resolution is likely to be even more politicized. As an active participant on behalf of incumbent Bell companies in these bitterly contested proceedings, as well as in proceedings under Section 271 to determine whether they had opened local markets to competition sufficiently to justify lifting the prohibition on interLATA service, I feel a powerful challenge to step back and reevaluate my views.

A Digression on the TELRIC Pricing Issues

My indignant opposition to some of the FCC's most important determinations—indignation assuaged only partially by the endorsement, adoption, and improvement by Justice Breyer in two separate Supreme Court decisions, 1999 and 2002[72]—was intensified, I must confess, by its failure to do justice to my own arguments; I pointed out in advance that the first volume of my *Economics of Regulation*, published in 1970, had anticipated the very error the FCC was about to commit.[73] Moreover, in so doing the commission also failed to take account of a separate but wholly coincident argument offered by Jerry Hausman.[74]

Hausman and I argued that with perfect competition, even if it were feasible, prices in an industry characterized by rapidly improving technology would not coincide with the average total costs of production using the most efficient new capital, as they would ordinarily be measured. Rather, investors would anticipate continuing reductions in costs linked to ongoing technological progress and therefore practice "anticipatory retardation," adopting the most recent technology only when market prices were sufficiently high to enable them to recoup a disproportionately large portion of their capital costs in the early years.[75] In a separate, striking chain of reasoning, Hausman concluded that the rental charges for such unbundled facilities would have to incorporate rates of return two to three times higher than typically permitted to regulated public utility companies, in order to compensate an ILEC for giving potential competitors the option of leasing on short term rather than purchasing the most modern facilities—a circumstance that would leave it to bear the costs of obsolescence of its assets.

Timothy Tardiff has graphically illustrated this principle by contrasting the very high cost of the short-term lease of a computer with the much lower cost of purchasing one on installment. Lessors demand higher compensation because they, rather than the lessee, bear the predicable costs of obsolescence.[76]

The FCC dismissed these arguments with the observation—entirely correct in principle—that its TELRIC formula could in fact allow ILECs to recover the anticipatable costs of obsolescence with appropriately higher rates of return. This rejoinder was explicitly endorsed by the Supreme Court decision sustaining the FCC's TELRIC formula, as well as by Rosston and Noll.[77] The reality—glossed over both by the Court and by Rosston and Noll—remains that when regulators actually got around to prescribing them, those rates bore an extremely close relationship to traditionally prescribed returns based on historical costs.

Temptations of the Kleptocrats

I continue to regard the FCC's decisions as acts of "astounding regulatory presumption."[78] It has also become increasingly clear how blatantly telecommunications deregulators at both state and federal levels have succumbed to the "temptations of the kleptocrats"[79]—by which I originally had in mind the temptation, notably in the case of electric power, to renege on obligations to allow utilities to recover their historical costs when those obligations stood in the way of politically expedient rate reductions.[80] In the case of the telephone industry, the FCC has invited state commissions to undermine rate caps that the majority had imposed on the telephone companies in order to give them financial incentives to improve productivity. Some of the prices prescribed under the TELRIC formula have implied immediate reductions in carriers' costs that it would have taken over twenty years to reach under the rate caps.[81]

The recognition by most state commissions that electric utilities had a right to recover stranded costs, more precisely, costs that would otherwise be stranded by deregulation,[82] understandably increased the willingness of the incumbent companies to cooperate in restructuring the industry. In contrast, the FCC's TELRIC prescription invited—indeed, demanded—obstruction on the part of the local telephone companies.[83]

The "UNE-Platform"

The hostility of the local telecommunications companies was further intensified, and the determination of the FCC to produce visible "results" further demonstrated, when the FCC confirmed that the ILECs must use the TELRIC benchmark to set prices for UNE-Ps (unbundled element network platforms), the entire combination of network elements (switches and transport facilities) necessary to produce the services. Competitors would then have the right to market the services to users under their own brands, effectively threatening to cut off the ILECs' access to the retail market.

It is only fair to disclose that my first objection to the UNE-P was aesthetic: "unbundled network elements" combined into a single bundle is an oxymoron. More substantively, the FCC's blatant invitation to the incumbent local companies to contest it every step of the way[84]—consistent with satisfying the commission that they were meeting the conditions allowing them to provide interLATA service—was especially ironic: it was certain to undermine the willingness of competitors to construct their own facilities.[85] Paradoxically, the consequence of the rapid increase in the use of UNE-Ps, beginning three years after passage of the 1996 act, was to *reduce* the number of lines served by competitive local telephone carriers with their own facilities.[86] Small wonder.

Brilliance in Hindsight

At first blush, there would seem to be no logical way in which the deliberate underpricing of UNEs dictated by the FCC can be blamed for what have proved to be wildly excessive investments in local facilities by CLECs. Had the commission been less generous to the CLECs—had its prescribed TELRIC pricing been less arrogant, less blatant in its denial of the ILECs' right to recover sunk costs, or at the very least, of their own incremental costs rather than the hypothetical ones prescribed by the FCC—what proves to have been massive overinvestment by CLECs in their own facilities would surely have been even greater.[87]

This Pickwickian exoneration of the FCC's policies would, however, ignore their timing and their self-contradictory nature—the commission's desire to encourage facilities-based local entry, on the one side, and its approval of UNE-Ps at TELRIC prices, on the other.[88] Investments by CLECs constituted 43 percent of the industry total in 1999–2000, before availability of the far less risky UNE-Ps became widespread. They declined sharply thereafter, as recourse to UNE-Ps soared[89]—at the expense of the ILECs—but even more catastrophically for entrants who had previously been foolish enough to construct their own facilities.[90]

Improvement in Foresight Accompanied by Intensified Myopia, or Worse

There were two important respects, however, in which the UNE-P surely does not deserve even this backhanded credit—for discouraging investment that proved shortly thereafter to have been excessive. In terms of the overall goal of encouraging competition, the first was its obvious discouragement of investment by competitors even in their own switches. Switches are at the heart of both product- and process-innovation in telecommunications. But what the UNE-P serendipitously minimized, preponderantly, was what proved after the fact to have been overinvestment in transport facilities: services using the switches of the incumbents, as well as their access lines, effectively constituted no competition, since all they did was cannibalize sales at retail of services actually provided by the incumbents. As Chairman Michael Powell and Commissioner Kathleen Q. Abernathy put it in their vehement dissent from the commission's long-delayed triennial review decision announced on February 20, 2003, the UNE-P produces the semblance of competition but not the substance:

> Consistently underlying my preferences in this area is a commitment to promote and advance . . . facilities-based competition that is meaningful and sustainable, and that will eventually achieve Congress' stated goal of reducing regulation. The benefits of such a policy are straightforward: Facilities-based competition means a competitor can offer real differentiated service to consumers—*the switch is the brains of one's network and to be without one is to be a competitor on life support fed by a hostile host.*[91]

Second, and most fundamental, the recovery of this devastated industry and the establishment of a genuinely competitive market depends upon facilities-based competition embodying technological innovation, rather than the hothouse competition induced by regulatory subsidization. From this standpoint, the decision of the bitterly split commission in February 2003, at the end of its second triennial review, is in two respects an important step forward, and in a third, an abomination.

The first important positive step—my explication of which can be brief because the case for it has been thoroughly expounded elsewhere[92]—was its 3-2 vote to exempt ILECs' investments in extending fiber to the home or neighborhood from both price regulation and the obligation to share with competitors.[93] These investments, which typically compete head-on with cable TV and wireless companies that are not subject to FCC restrictions, would surely be discouraged by mandatory sharing—particularly at the FCC's prescribed hypothetical TELRIC costs.[94]

The other positive step was the effective endorsement of the Powell/Abernathy emphasis on the importance of competitive provision of switches: all the commissioners agreed on withdrawing the UNE-P from large business customers—specifically, customers served by high-capacity loops.

In a third respect, however, the commission's decision was grounded in neither good economics nor honorable regulatory practice. This was the 3-2 vote (with both the chairman and Commissioner Abernathy vigorously dissenting) to surrender to the states the FCC's statutory responsibility for defining the unbundled network elements of the incumbent companies' systems that they are required to lease to would-be competitors—including the authority to continue the UNE-P requirement—for serving residential and small business customers. Six years after the commission required incumbents to unbundle essentially every element of their networks—for which it was rebuked first by the Supreme Court and later by the D.C. Circuit Court of Appeals[95]—at ridiculously low rates, a majority of commissioners has now decided to transfer back to the state commissions full and *unreviewable* authority over the prescription of inputs for residential and small business services subject to mandatory sharing—*including the critical switches.*[96] The one certainty is that that devolution will turn out to have resolved little or nothing.[97] The courts, not the commission, will have the final word on the question of "unreviewability." This part of the decision guarantees years and years of additional litigation and regulatory uncertainty.[98]

The state commissions are, of course, under even more direct political pressures than the federal agency to produce "results"— visible, even if only paper, competitors and visible reductions in residential rates.

Triumphs of the Kleptocrats

We have already had a foretaste of the results that the new, politically minded FCC majority hopes to achieve, as local incumbent companies lost millions of subscribers within a space of months to free-riding competitors such as AT&T and WorldCom. Last year the Michigan commission cut the state's wholesale rates from about $17.50 to $14.44 per line per month. "I really don't care what form the competition takes, so long as companies are in there duking it out and fighting for customers," the chairman was quoted as saying.[99] Some competitors! Some "duking"!

Similarly, in May 2002 the California commission ordered a 39 percent reduction (from $23 to $14 a line) in combined charges for loops and switching at high levels of usage. Also in early 2002, under pressure from the New York Public Service Commission, Verizon entered into a settlement that had the effect of reducing its UNE-P rate from $27.17 per line to $19.14, while the District of Columbia commission reduced the line charge to around $5.

I have had direct exposure to an analysis of one of these public utility commission–prescribed downward adjustments in the price for UNEs (both UNE-P and UNE lines alone)—this one by the Illinois Commerce Commission for SBC Illinois. Beginning by pointing out that the new rates are either the very lowest among the fifty states or very close, Debra Aron demonstrated to my astonishment that they "do not even generate sufficient revenues [for SBC] to cover its day-to-day cash expenditures to provide those UNEs, even excluding the cash demands for taxes, bondholders, and investors."[100] As I testified at SBC's request, I could not as a regulator have possibly justified setting any rates unless they were explicitly subsidized by others at such a level except if I had made some explicit finding that a company's management was deeply negligent.[101]

The apparent total failure of the Illinois Commission to reconcile the fill factors it employed to calculate SBC's TELRICs—which were far higher than the fill factors the company had actually been achieving—with the fact that *it had adopted rate cap regulation eight years earlier* was equally astounding. The case for rate caps was, of course, that they would provide incentives for efficient operation much stronger than traditional cost-plus regulations—as further assurance, they incorporated a stipulated 4.3 percent annual reduction in real terms. If the UNE rates set by the Illinois commission do indeed imply a reduction in SBC-Illinois' costs that, according to Aron, would take at least twenty years to achieve under its own rate caps, there is surely a case to be made here for legislative if not *judicial* intervention to enforce a presumption that rates issuing from that reform would be just and reasonable.[102]

In ordinary circumstances I would not regard it proper for the legislature or the judiciary to encroach upon the presumed superior expertise of an independent regulatory agency. When, however, a state commission interprets a new mandate from the federal government in a way that yields so dramatic a departure from the rates issuing from its own reforms adopted many years previously, it seems to be entirely proper for the courts or the state legislature to call it to account.[103]

I observed, also, that the provisions for depreciation and obsolescence in the Illinois Commerce Commission's newly prescribed rates appear not to have reflected the assurances by the FCC in approving the TELRIC method—assurances later accepted by the U.S. Supreme Court—that TELRIC should be calculated in a way that compensates for the unusually high rates of obsolescence in as technologically progressive an industry as telecommunications. Instead, the Illinois commission simply adopted the depreciation rates prescribed by the FCC *before* the 1996 federal act proclaimed the national policy of competition, *before* the technological change the policy was intended to accelerate, and *before the FCC had prescribed its genuinely radical TELRIC formula*. In these circumstances, too, it seems entirely proper for the state legislature to have called the Illinois Commerce Commission to account—even to the point of prescribing a depreciation formula developed outside the context of regulatory rate setting.[104]

SBC has estimated that the two changes directed by the new law would increase its monthly UNE-P charges from $12.38 to about $21.48. Indeed, its passage was bitterly denounced by consumer representatives and retail competitors (most prominently, AT&T and MCI) as threatening to squeeze out competition and to produce outrageous increases in local residential rates. Yet—paradoxically, it would appear—this statute decrees that "unbundled network element rates established in accordance with the provisions of this Section shall not require any increase in any retail rates for any telecommunications service."

I could not, of course, predict with total confidence that some retail rates offered by SBC's competitors would not have to go up: if AT&T and MCI were able to offer residential customers reduced rates only because they could purchase the services from SBC bundled at a price set far below SBC's own avoided costs, clearly they would have to raise them. They would almost certainly also have to accept narrower profit margins—a not unreasonable prospect in light of the fact, reported by AT&T to the FCC, that competitive carriers in Illinois can currently earn the highest gross margins in the country.[105]

The fact is, however, that local regulated telephone companies have always been required by state regulatory agencies to hold their basic residential rates below book costs, compensating by charging far above cost for most of their other services, such as intrastate long distance, access to their networks by other long-distance companies, local service to large businesses, and service features such as call-waiting and caller ID. If the new local competitors are made to pay SBC Illinois its true costs—and no more—for the network elements they lease, there is no reason why they cannot match SBC's regulated basic charges, making up any deficiencies in those revenues by marketing additional high margin services to residential subscribers. And that is exactly what I understand they were doing.

If competitors such as AT&T and MCI are as efficient as SBC, there is no reason why they cannot compete with it when they are charged wholesale prices equated to its costs. If they are not as efficient, they do not deserve to survive. Subsidizing competitors at the expense of incumbents is a cheap way of getting political credit, but it is not a way of encouraging efficient competition—or, in the long run, of promoting consumer welfare.[106]

Three years ago I published an article titled "Bribing Customers to Leave and Calling It 'Competition.'"[107] The title refers to the policy, adopted by some states, of forcing local electric utility companies to give rebates to retail customers who deserted them for competitive distributors—that is, the portions of the distribution charges that customers would escape if they left their historical retail suppliers—larger than the costs that their departure would actually save the incumbents. In effect, their commissions reasoned: "we estimate that customers who desert their local utility suppliers will save it, say, 3.5 cents a kwh—the cost of the energy it will no longer have to purchase in the wholesale market in order to supply them—but we will make the company give them a 'shopping credit' of 4.5 cents, in order to encourage them to shift." The Pennsylvania commission, to cite the outstanding example to date, deliberately prescribed a "shopping credit" large enough to produce something like a 10 percent rate reduction for customers who shifted to competitive marketers. One of its commissioners subsequently boasted that, as a result, more customers had shifted in that state than in the entire remainder of the country.

Now the same thing is happening in the telephone industry, and a majority of FCC commissioners have left the states free to continue the practice. They thereby earned praise from the *New York Times*, on the ground that "consumers are only now beginning to benefit" from "competition."[108] The regulatory agency sets not the price of a network element or elements "essential" or "necessary" for competitors, but the wholesale price of the *service* or bundle of services, admittedly below the actual or avoided costs of the incumbent. *Some* competition! *Some* deregulation!

This reaffirmation of positions I have taken in adversarial proceedings on behalf of the local Bell companies must be posed against the view, at least equally prevalent, that the incumbent local companies are the villains of the process and, by continuing to refuse to cooperate wholeheartedly in opening their systems to competitors, they are the principal obstacle to deregulation. For example, in an article on "the great telecoms crash," the *Economist* predicts:

> The likely winners, it is already clear, are the former "Baby Bells" in America and the former monopoly incumbents in Europe. Because they own the "last mile" of the network that runs into homes and offices, these operators have a firm grip on their customers and solid revenues. Compared with their upstart competitors that proliferated after the liberalisation of telecoms

markets during the 1990s, these firms are relative safe havens. Customers can switch long-distance carriers at the first whiff of trouble, but often have no choice of local provider. In theory, regulators should require local monopolies to allow competitors to provide services over their lines, but most local monopolies have successfully obstructed such "local-loop unbundling" using a variety of technical excuses.[109]

The Baby Bells' emergence as "winners" did not require systematic obstruction of competitors for the historical monopoly providers of what society generally regards as a basic necessity of life to have emerged from the speculative boom more successfully than the hundreds of imitators attracted by government policies designed to stimulate entry by "competitors." The market value of major local Bell companies—Verizon, SBC, and Bell South—was, as of early October 2002, down more than 40 percent from its peak.[110] This decline was hardly surprising in view of the very rapid growth in non-facilities-based competitors, which were given access to the UNE-Ps at prices markedly below the ILECs' own incremental costs. As for the assertions that the Bells have systematically obstructed local competitors, it seems to me there is no way an outsider—or most insiders, for that matter—can evaluate the conflicting complaints and defenses. An independent assessment would have to take into account the ILECs' years of intense effort, at the cost of billions of dollars collectively, to convince highly skeptical federal and state regulators that they were indeed cooperating in opening their local markets sufficiently to satisfy that precondition for their offering interLATA services set by the 1996 act. Moreover, historical experience clearly demonstrated the compatibility of the ILECs' positions and policies with the enormously successful growth of competitive wireless service, which was equally dependent upon access to their facilities.[111] Then, too, one can point to the extraordinary growth of the facilities-based CLECs themselves before the meltdown—facilities that will, of course, not disappear even if the CLECs do—during these periods of claimed obstruction.

A Reentry of Antitrust?

The *Trinko* case raises the question of whether the ILECs' compliance with the requirements of Sections 251 and 271 of the act—compliance already subject to exhaustive administrative rules and sanctions, including withholding or retraction of authority to serve the interLATA market—should also be subject to suits by competitors under the antitrust laws.[112] Here I can only plead schizophrenia. For what it is worth, I set forth the major contributors to that condition:

First, my consistent subscription to what I take to be

> [t]he predilection of most economists . . . to entrust the responsibility for managing the transition from regulation to deregulation to the antitrust authorities, with their presumed superior expertise in comprehending the requirements of effective competition, rather than to the former regulatory agencies, particularly because of the demonstrated proclivities of those agencies to protectionism and cartelization, including a tendency to condemn any and all price-cutting as "predatory" or "destructive."[113] Closely related are the tendencies of regulators to seize opportunities to produce reductions in the rates for still-regulated services and visible *competitors*, at the expense of competition, in the unregulated markets.[114]

Next, my recognition, as a long-time participant in these bitterly contested regulatory proceedings, that

> I cannot yet bring myself to deny the regulatory agencies a central role in the transition of public utilities to competition, in view of the special

circumstances of those industries: the necessity for settling out and tracking the collection of strandable costs; the more pervasive possibilities in those industries of tying competitive to monopoly services, directly or subtly, and of cross-subsidization, strictly defined; the consequent need for accounting separations and the monitoring of transactions between still-regulated utilities and unregulated affiliates; and, finally, the pervasiveness of essential facilities controlled by incumbents—not to mention competitive advantages deriving solely from their historical franchised monopolies, requiring an administrative agency to define them and to prescribe the terms and conditions of sharing.[115]

Finally, my initial reaction upon hearing of the *Trinko* case:

Oh my God! I have spent the last seven years (at least) involved in endless administrative proceedings under Sections 251 and 271 of the Telecommunications Act and under state regulatory statutes before that—mainly on behalf of Baby Bells—in the latter of which the Bells had to demonstrate to the FCC (with the Department of Justice playing an advisory role) compliance with the highly detailed provisions requiring demonstration that they had opened their local markets to competition—accommodations costing, as I recall, billions of dollars; and under which (as once again I recall) the CLECs and would-be CLECs retained the right to complain to the public utility commissions of asserted acts of noncompliance (such as alleged refusal of ILECs to permit employees of CLECs located in their exchange offices to use the bathroom)—the notion of re-litigating those cases before juries gives me nightmares. [I might well have added that the act imposes on the ILECs obligations to positively assist competitors going far beyond even the most liberal interpretation of the essential facilities doctrine in antitrust.] And yet I have no answer to the principle that assertedly injured parties must not be denied remedy under the antitrust laws—unless Congress may reasonably be interpreted to have intended to substitute the administrative arrangements for antitrust.[116]

Line Sharing

An intensely contested issue in the last several years, which may or may not prove to have been resolved by the FCC's decision of February 2003, has been whether the ILECs should be obliged to permit competitors to use the high-frequency capability of their ubiquitous copper customer-access lines. Perhaps as compensation to the ILECs for returning them to the mercies of the state commissions on whether they should be required to offer UNE-Ps to residential customers and at what price, perhaps as a consistent accompaniment to relieving them of the obligation to lease to their competitors any investments they make in the installation of high-speed fiber that can deliver broadband service, the commission's recent 3-2 decision rescinds that line-sharing obligation. The dissents of Chairman Powell and Commissioner Abernathy remind me of my recognition at the time of the FCC's original decision that this capacity of the ILECs' copper networks, which was exploitable at something close to zero marginal production cost, would seem to be the archetypal case for mandatory sharing.[117] The copper wires, after all, were inherited from their franchised monopolies, the sharing of which facilities would not discourage future risky investments by the ILECs. I was eventually persuaded to oppose that mandate by the FCC's own reasoning elsewhere that the broadband market was already effectively competitive because of the availability of cable and wireless. Hence, the commission concluded that entrants' access to the incumbents' copper lines could no longer be said to be necessary to competition. Nor could it be reconciled with the fact that the cable companies had no equivalent sharing obligation.

The close-to-zero marginal cost of this particular form of access, however, would still appear to offer a very large competitive advantage to the ILECs that was not the product of superior foresight or risky, costly innovation, and might therefore legitimately continue to be subject to mandatory sharing. True, making that capability of their copper lines available to CLECs at incremental costs would dilute the incentives of those competitors to construct their own broadband transport facilities. But it would do so no more than the difference in the true incremental costs of the two options would dictate: here, the FCC did not prescribe its hypothetical blank-slate price for line sharing. True, too, forced sharing of this capacity would conceivably have high *opportunity* costs for the ILECs, in the sense that it would exclude them from use of those capabilities of their own facilities.[118] But a sharing mandate for high-frequency capacity would be no more intrusive than the statute's requirement that ILECs lease low-frequency voice capabilities when a CLEC has competed successfully for a subscriber—a requirement to which, so far as I know, no ILEC has objected in principle.

And, finally, the line-sharing requirement did, in fact, reflect its low incremental production cost and did generate genuine price competition in the offer of broadband—a source of competition that is now threatened by the commission's February 2003 about-face.[119]

My continuing uncertainty, to which I reluctantly confess, springs from

—the distorting effect of the imposition of mandatory line sharing on the ILECs but not on the cable companies, which have doubled their market share and enjoy the parallel advantages of inheriting broadband capability from *their* franchised monopolies;[120]

—the fact that adding DSL (digital subscriber line) capability to their copper lines was and continues to be both costly and commercially risky, as well as subject to intense competition from cable;[121]

—the reality that continued occupation of the low-frequency portion of the copper wire still carries with it the ILECs' obligation to offer basic service at regulated rates, while line sharing deprives them of the opportunity to compensate by offering broadband service with the same lines; and

     —that unregulated cable telephony, using voice-over-Internet protocol (VOIP) and subject to no sharing requirements, is emerging as a real competitive threat to traditional telephony, both local and long distance.[122]

4 | *Conclusion*

The meltdowns in commercial aviation and telecommunications have been catastrophic to almost all the parties financially involved. But they offer no valid basis for reversing deregulation in these two sectors. On the contrary, the failures have been failures of continued governmental attempts to manage the process in such a way as to produce quick, politically popular results, as distinguished from simply freeing competition—and especially technological competition—to work its will.

Deregulation shifts the major burden of consumer protection to the competitive market, and therefore, in important measure, to the enforcement of the antitrust laws. But the experiences with essentially unmanaged deregulation in airlines and pervasively managed deregulation in telecommunications also demonstrate that the focus of policy should be, first and foremost, on liberating competition from direct governmental restraint—not on dictating market structures or outcomes. In particular, while in no way counseling indulgent

47

antitrust treatment of predatory or unfairly exclusionary competitive *conduct*, they underline the need for humility in attempting to make industries more competitive by interfering with their achievement and exploitation of economies of scope and diluting their incentives for innovation—the most powerful competition of all.

Notes

1. The total market capitalization of all U.S. airlines (that is, the seventeen listed in the Bloomberg U.S. Airlines Index as of December 3, 2002, plus TWA and US Airways) was $60,372 million in mid-July 1998. It had dropped to $34,451 million by September 10, 2001, and subsequently fell to a level of $20,536 in early December 2002. By the end of May 2003 it had recovered slightly, to $21,998 million. Excluding Southwest and Jet-Blue, however, the airlines' market capitalization was $53,285 million in mid-July 1998 but had plummeted to $6,717 by early December 2002 and rose only to $7,317 million by the end of May 2003. Data from FactSet Research Systems and Bloomberg LP.

As for the telecommunications industry, the total market value of *competitive* local exchange carriers dropped from a high of some $85 billion to $4 billion, "at last count." Competitive Enterprise Institute, *Weekly Commentary*, January 31, 2003; see also Larry F. Darby, Jeffrey A. Eisenach, and Joseph S. Kraemer, *The CLEC Experiment: Anatomy of a Meltdown* (Washington: Progress and Freedom Foundation, September 23, 2002). Lucent and Nortel, the two dominant suppliers of switches (and heirs to Western Electric), suffered a 99 percent loss in market value; see Thomas W. Hazlett, "The Irony of Regulated Competition in Telecommunications," *Columbia Science and Technology Law Review*, vol. 4 (2003), p. 8. The local Bell telephone companies faced "only" a 40 percent decline; on this and their partial subsequent recovery, see note 110, below, and the associated text.

2. Similar questions are, of course, being asked about the restructuring of the market for electric power, mainly because of the catastrophic experience of California. I make no attempt to answer them here, for a very simple reason—cowardice.

3. Yet, significantly, the most bitter public criticisms come not from industry or financial but from populist circles: "Deregulation Was Supposed to Cut Prices, Expand Choice, Enhance Service—Improve Your Life. So How Come You're Not Smiling?" *Consumer Reports* (July 2002), pp. 30–35 (something less than one full page, however, is devoted to the airline experience).

4. This is the reason for the title of my book *Letting Go: Deregulating the Process of Deregulation* (Michigan State University Institute of Public Utilities, July 1998).

5. J. Gregory Sidak, "The Failure of Good Intentions: The WorldCom Fraud and the Collapse of American Telecommunications after Deregulation," *Yale Journal on Regulation*, vol. 20, no. 2 (Summer 2002), p. 213. His figure 3, along with a telephone call to the Federal Communications Bar Association on June 3, 2003, demonstrate that the membership has remained at its 1998 level—small wonder.

6. The NASDAQ, home to many of these stocks, fell by 78 percent from its all-time high of 5048.6 on March 10, 2000, to 1114.1, its lowest level, on October 9, 2002. By May 30, 2003, it had recovered to 1595.9.

7. There is yet another separate story to be told about the responsibility of the protracted and messy process of the deregulation of the electric utility industry for the "credit crisis" confronting and afflicting it now; see "Electric Industry Hits Credit Crisis," *Wall Street Journal*, October 15, 2002, p. 2A. Many of the critical issues associated with this crisis remain to be resolved; but just as there were common elements in the "deregulation" process, so also there are major respects in which each industry—aviation and telecommunications—and the process of its deregulation (or "restructuring," for electric power—observe the significance of the difference between those two concepts) make me lack confidence in my ability to offer nuanced generalizations reasonably applicable to each and all.

8. Sam Peltzman and Clifford Winston, eds., *Deregulation of Network Industries: What's Next?* (AEI-Brookings Joint Center for Regulatory Studies, 2000), p. 2.

9. For a recent poignant demonstration, observe the failure of American Airlines' "More Room in Coach" marketing plan; see Edward Wong, "American Air Is Adding Seats and Cutting Legroom in Coach," *New York Times*, May 22, 2003, p. C4.

10. It is worth underscoring my assumption here, that by removing price floors, deregulation has indeed exacerbated the financial losses inflicted on the industry by recessions. It has also, however, given carriers greater freedom than they would have otherwise had to mitigate the effects of recession—for example, by deserting unprofitable markets and curtailing unprofitable operations. It is worth pointing out, therefore, that Steven Morrison and Clifford Winston have tested the assumption, by comparing how the airlines actually reacted with how they would have behaved had the pre-1978 regulatory regime been maintained. They conclude that deregulation did enable the carriers to fare better financially. See Steven Morrison and Clifford Winston, *The Evolution of the Airline Industry* (Brookings, 1995), pp. 103–05.

Domestic load factors have remained surprisingly stable at around 71 percent through the traumatic events of 2001, 2002, and the first months of 2003. The carriers managed this only by mothballing aircraft (but not, alas, the associated fixed costs) and at the expense of average yields, which dropped from 14.49 cents in 2000 to 13.25 and 12.04 cents, respectively, in the next two years. Rates of return on invested capital, of course, turned abruptly negative. Data from Air Transport Association Office of Economics, *Monthly Passenger Revenue Report* (Washington, various issues).

11. Richard Branson has recently repeated his desire—first expressed several years ago, but frustrated by the 25 percent limit—to set up an American version of his successful low-fare Virgin Airlines.

12. Boaz Moselle and others, *The Economic Impact of an EU-US Open Aviation Area*, report prepared for the European Commission Directorate-General Energy and Transport by the Brattle Group (London and Washington, December 2002); also Scott McCartney, "U.S. Airlines Would Benefit from Foreign Competition," *Wall Street Journal*, January 22, 2003.

13. Visualize an airline with eighteen planes, nine lined up from north to south along each coast and ready to fly across the continent. Without a hub somewhere in the middle, all it can offer in any one bank of flights is eighteen origins and destinations—one for each aircraft. With a single hub somewhere in between, those eighteen aircraft can service 198 city pairs going in the same direction, and another 144 for passengers doubling back to destinations on the coasts on which they started. Count them. Bear in mind, too, that the original justification for code-sharing alliances among airlines of different nationalities was to enable their customers to reap the same benefits of hubbing (and competition between hubs) in the international arena.

14. Even readers considerably less sensitive than I to the implications of my having here added the qualifying "seemingly" to my previous indignant characterizations of these fares might find an explanation of interest. I offered those unqualified denunciations of some fares at the time when the industry was earning extremely satisfactory profits overall and, as I also frequently emphasized, the fruits of monopoly took the form of clearly higher-than-competitive wages for large segments of airline employees. The abysmal present financial condition of American Airlines, US Airways, and United Airlines, along with my long-standing recognition (going back to my reading of J. M. Clark's *Studies in the Economics of Overhead Costs* [University of Chicago Press, 1923]) of the necessity—indeed, within limits, benevolence—of price discrimination in industries with heavy fixed costs, clearly have modified—but have hardly eliminated—my outrage at the $841.50 Ithaca-to-Washington roundtrip fare that I had to pay earlier this year to testify at a Senate hearing.

15. Micheline Maynard, "A Premium Summer for Low-Fare Airlines," *New York Times*, June 29, 2003, Travel Section, p. 3.

16. The defense I offered in 1978 of my publicly professed inability to predict the future structure of a deregulated industry seems to me just as valid today as it was then: "So far as I know there is no objective basis for deciding which of these situations is more likely to prove typical—the one in which the integrated, typically larger carrier can out-compete its more specialized rivals because of the network economies it enjoys from the multi-stop and on-line traffic flows at its command, or the one in which the specialized carrier will have clear advantages. . . . Perhaps the only conclusion

one can and need draw is that under a competitive regime, where all suppliers are free to try to take advantage of the opportunities presented by free entry and exit to augment their advantages and shed their handicaps, and to have their qualifications tested in the marketplace, these various kinds of market situations can be expected to sift themselves out automatically, with various kinds of suppliers emerging successful on the basis of their respective advantages and handicaps in each. I admit to the slightest feeling that this conclusion may be just a little glib, but it seems to me the only one justified by the experience with market competition in this industry and in the economy at large.

"Moreover, if we cannot *predict* how these offsetting advantages and handicaps of the several carriers are likely to work out under a regime of free entry, it seems to me even less likely that we can hope to achieve the most efficient performance of the transportation function by *prescribing* how the thousands of markets should be served, through a comprehensive regulation of entry. I find it difficult to see how these uncertainties tilt the balance in the direction of a reliance on ignorant regulation in preference to an uncertainly predictable market process." (Alfred E. Kahn, Talk to the New York Society of Security Analysts, New York City, February 2, 1978.)

17. The fact that the recourse to such combinations has largely been motivated by financial catastrophes must not be permitted to obscure the fact that consolidation has long been driven by efforts to reap the benefits of hub-and-spoke operations over wider geographic areas. See also note 23, below. On the negative side, such combinations have always entailed a threat of suppressing head-to-head competition between the partners or putting nonmembers at an unfair competitive disadvantage by denying them an equal opportunity to interchange traffic. This raises the possibility that consolidation will lead the U.S. industry to the sort of oligopolistic structure that dominates the international field. See Committee for a Study of Competition in the U.S. Airline Industry, *Entry and Competition in the U.S. Airline Industry: Issues and Opportunities*, Special Report 255 (Washington: National Research Council/Transportation Research Board, 1999), pp. 145, 150–51.

The insistence of the British government on conditioning its entry into an open skies agreement with the United States on our approval of the long-proposed alliance between American Airlines and British Airways led me to come out in its support in 2001, despite the fact that the two companies were and are direct competitors on some—most importantly, United States to Heathrow—routes. But the deal was apparently insufficiently attractive to the U.S. Department of Transportation unless it were accompanied by large divestitures of the two airlines' slots at Heathrow Airport—a condition rejected by the British government. See my October 30, 2001, statement in that proceeding, pointing out that the increase in concentration at Heathrow would be minimal and that the argument about concentration and redistribution of Heathrow slots was an argument about the distribution of economic rents, rather than the effectiveness of competition ("Statement of Professor Alfred E. Kahn," American Airlines, Inc. and British Airways PLC—Antitrust Immunity Agreement, U.S. Department of Transportation Docket nos. OST-2001-10387 and 10388, November 2001; available at http://dms.dot.gov/search/document.cfm?documentid=142794&docketid=10387). While welcome, the Department of Transportation's ultimate approval in part of that code-sharing agreement does not cover nonstop service between the United States and London or grant the antitrust immunity required to permit the full integration of their offerings—and is not accompanied by an open-skies agreement ("DOT Announces Final Approval of American-British Airways Code-Sharing," available at www.dot.gov/affairs/dot4603.htm).

18. The danger is that the alliance would have the incentive and ability, by virtue of its agreement among the members, to reduce competition on more city pairs than would be called for by efficient, competitive service of the (by assumption, secularly declining) demand. See Michael E. Levine, "Airline Competition in Deregulated Markets: Theory, Firm Strategy, and Public Policy," *Yale Journal on Regulation*, vol. 4, no. 2 (1987), pp. 393–494; Severin Borenstein, "Hubs and High Fares: Airport Dominance and Market Power in the U.S. Airline Industry," *Rand Journal of Economics*, vol. 20 (Autumn 1989), pp. 344–65; Severin Borenstein, "Airline Mergers, Airport Dominance, and Market Power," *American Economic Review*, vol. 80 (May 1990), pp. 400–04; Alfred E. Kahn, "The Competitive Consequences of Hub Dominance: A Case Study," *Review of Industrial Organization*, vol. 8, no. 4 (1993), pp. 381–405.

19. For this reason the three carriers said they would feel compelled to pursue a similar merger if the UAL–US Airways union were permitted.

20. On the possible anticompetitive consequences of the proposed alliance of Delta, Continental, and Northwest Airlines, see Jonathan B.

Baker, Albert A. Foer, and Alfred Kahn, "Proposed Air Alliance of Delta/Continental/Northwest Raises Important Antitrust Questions," Joint letter to Transportation Secretary Norman Y. Mineta and the Department of Justice's Charles James, November 12, 2002 (www.antitrustinstitute.org).

21. Alfred E. Kahn, "Statement on the State of Competition in the Airline Industry," before the U.S. House of Representatives Committee on the Judiciary, June 14, 2000, p. 8 (www.house.gov): "The likelihood that a United/US Airways merger would indeed result in suppression of such potential competition would seem to be enhanced by what I take it would be United's explanation and justification—namely, its need for a strong hub in the Northeast (commented on widely in the literature, along with attributions of a similar need to American Airlines). But if United really does feel the need for a big hub in the Northeast, this suggests that it is indeed an important *potential* competitor of US Airways, and that, denied the ability to acquire the hub in the easiest, noncompetitive fashion, by acquisition of that company's Pittsburgh and Charlotte hubs, might instead feel impelled to construct a hub of its own in direct competition with it."

22. Skepticism seems especially compelling in cases where an airport is already dominated by one or another of the parties—Minneapolis–Saint Paul, Detroit, Houston, Cleveland, Newark, Salt Lake City, Cincinnati, and Atlanta. If the greater majority of flights coming in to and going out of these airports are the flights of a single carrier, it is difficult to see how the gates, ticket counters, or baggage handling facilities used by its new partner can physically be brought closer together than they already are.

23. This has of course been particularly true in the international arena, where carriers typically are not free to enter whatever markets they choose and establish hubs of their own.

24. "[W]e find that average fares fell and total traffic increased between cities that were served by the alliances after the alliances began . . . online fares on a one-stop itinerary typically are lower than the sum of fares on the two flights making up an interline flight, online fares typically are lower than interline." Gustavo E. Bamberger, Dennis W. Carlton, and Lynette R. Neumann, "An Empirical Investigation of the Competitive Effects of Domestic Airline Alliances," Working Paper 8197 (Cambridge, Mass.: National Bureau of Economic Research, March 2001), pp. 2, 4. See also Department of Transportation, *International Aviation Developments: Global Deregulation Takes Off,* First Report (December 1999).

Dennis L. Weisman has cited this kind of internalization of demand externalities (in the form of call-back or reverse-to-offer elasticities) as a likely benefit of mergers of geographically separated local telephone companies. See my summary in Alfred E. Kahn, *Whom the Gods Would Destroy, or How Not to Deregulate* (AEI-Brookings Joint Center for Regulatory Studies, May 2001), p. 44, and Weisman's subsequent generalization of the argument in Dennis Weisman, "Market Concentration, Multi-Market Participation and Antitrust," working paper, Kansas State University, 2003; and Dennis Weisman, "A Generalized Pricing Rule for Multi-Market Cournot Oligopoly," *Economics Letters* (2003, forthcoming).

25. Jan K. Brueckner and W. Tom Whalen, "The Price Effects of International Airline Alliances," *Journal of Law and Economics*, vol. 43 (October 2000), pp. 503–45. See also Moselle and others, *The Economic Impact of an EU-US Open Aviation Area*, chap. 4.

26. Alfred E. Kahn, "Comments on Exclusionary Airline Pricing," *Journal of Air Transport Management*, vol. 5, no. 1 (1999), pp. 1–12; and Alfred E. Kahn, "The Deregulatory Tar Baby: The Precarious Balance between Regulation and Deregulation, 1970–2000 and Henceforward," *Journal of Regulatory Economics*, vol. 21, no. 1 (2002), pp. 35–56.

27. In suggesting that the House Committee on the Judiciary consider the differential effect of the proposed alliance on members and nonmembers, I recognize that I will almost certainly be accused of favoring "soft" over "hard" competition—that is, seeking to protect outsiders from disadvantages flowing from the limitations in their modes of operation and attractiveness of their services, by hampering their network competitors from exploiting advantages that flow from their in some ways superior mode of operations. This is a risk to which I have been exposing myself over at least a half century. See Joel B. Dirlam and Alfred E. Kahn, *Fair Competition, the Law and Economics of Antitrust Policy* (Cornell University Press, 1954). But when three competitors, constituting 35 percent of total business at the national level—and, of course, correspondingly much higher percentages in particular markets affected—agree to extend to one another, but not to their smaller challengers, the benefits of their combined economies of scope, it is impossible not to ask whether consumers are likely to be helped or hurt, on balance, and by what precautions nonintegrated competitors can be protected from exclusionary practices such as exclusive patronage refunds.

28. Michael E. Levine, "Re: Review of Proposed Alliance of Continental, Delta and Northwest Airlines," Letter to Transportation Secretary Norman Mineta, December 17, 2002.

29. Department of Justice, "Department of Justice Approves Northwest/Continental/Delta Marketing Alliance with Conditions," press release, January 17, 2003.

30. W. Tom Whalen argues persuasively that full achievement of these benefits, which he estimates as fares 19 percent lower than nonalliance fares, requires immunity from the antitrust laws and profit sharing—both explicitly forsworn by the Continental-Delta-Northwest alliance. W. Tom Whalen, "Constrained Contracting and Quasi-Mergers: Price Effects of Code Sharing and Antitrust Immunity in International Airline Alliances," Economic Analysis Group Discussion Paper (U.S. Department of Justice, Antitrust Division, April 24, 2003). Compare the much more limited approval of the American Airlines British Airways alliance, note 17, above.

Jonathan Baker, in personal correspondence, offered the intriguing conjecture that "the code-share alliance may induce tacit coordination among its members, leading the carriers to reduce segment fares by striking an implicit 'you lower yours and I'll lower mine' bargain"—a beneficent variation of the conscious parallelism ("conscious vertical complementarity"?) that in its original form represented, in effect, a frustration or circumvention of section I of the Sherman Act.

31. Department of Justice, "Department of Justice Approves . . . Marketing Alliance with Conditions," p. 5.

32. Ibid., p. 6.

33. See, in this connection, Jonathan B. Baker, "Mavericks, Mergers and Exclusion: Proving Coordinated Competitive Effects under the Antitrust Laws," *New York University Law Review*, vol. 77 (April 2002), pp. 135–203. Baker identifies Northwest Airlines as just such a "maverick." Concerning the trial brief of the United States opposing the projected, but then abandoned, merger between Continental and Northwest, as Baker puts it: "By refusing to match, Northwest and Continental have, at differing times in the past, each prevented price increases that the rest of the industry has tried to impose. Indeed just before this suit was filed, Northwest had frequently refused to match systemwide increases, including some increases that Continental initiated. As a result of Northwest's ownership of Continental, however, the two

carriers' interests are more closely aligned, making it less likely that either will be the one 'spoiler' that blocks a nationwide price increase. Fewer potential spoilers will lead to higher prices for consumers."

Ironically, Baker's conjectural conscious *vertical* complementarity that I describe in note 31, above, represents, in effect, an optimistic expectation that the fellow-members might indeed *not* price as independently as if they were not aligned in this way!

34. See references in note 17, above.

35. See the elaboration of this argument in Baker, Foer, and Kahn, "Proposed Air Alliance . . . Raises Important Antitrust Questions." In listing this feature of the alliance as a benefit to consumers on a par with its "offering codeshare service to new cities, increasing frequencies or improving connections," the Antitrust Division ignores the close analogy to tie-ins and full-line-forcing by agreement among respective hub-dominating competitors.

36. For a recent status report on the issue, with views congenial to mine, see Eleanor M. Fox, "What Is Harm to Competition? Exclusionary Practices and Anticompetitive Effect," *Antitrust Law Journal*, vol. 70, no. 2 (2002), pp. 371–411.

37. Kahn, *Whom the Gods Would Destroy*, pp. 39–40.

38. Ibid., pp. 39–45. On the subsequent compromise settlement, see Department of Transportation, Office of the Secretary, "Termination of Review of Joint Venture Agreements," March 31, 2003.

39. "[T]he Delta/Continental/Northwest alliance is not an end-to-end alliance, unlike most of the other domestic and international alliances reviewed by us, which typically have expanded the network of each alliance partner. . . . While Delta's code-share with Northwest would give Delta access to significantly more domestic airports, it appears that the total number of new on-line markets created by the alliance would still account for only 89,530 annual passengers, far less than one-tenth of one percent of all domestic passengers. Thus, the value of the alliance for the partners would come from capturing passengers now traveling on other airlines, rather than the stimulation of traffic in new 'online' markets. As a result, the proposed alliance would not provide substantial network extension benefits, unlike other domestic alliances." Department of Transportation, "Termination of Review under 49 U.S.C. § 41720 of Delta/Northwest/Continental Agree-

ments" (hereafter, "First DOT Opinion"), released January 17, 2003, unpaginated (pp. 7–8).

Some economists argue that the antitrust agencies should be required to weigh the possible benefits of mergers, particularly, against potential threats to competition. I have always counted myself a skeptic of the capability of the courts to make such assessments and, correspondingly, a proponent of the position that the laws should concentrate on the threat of the arrangements under scrutiny to the vitality of the competitive process. Arguably, regulatory agencies have the requisite greater expertise. But I incline to the position that the injunction should apply equally to them, to the extent they continue to have antitrust responsibilities.

40. See my general agreement with the department's effort to strengthen the protections against putatively predatory responses of incumbent hub carriers to the entry of low-fare competitors in "A Digression on Assertedly Predatory Airline Pricing" in Kahn, *Whom the Gods Would Destroy*, pp. 36–39 (text), and 67–70 (notes). I see no conflict between that position and the one that I express here, for the following reasons:

—The questionable pricing practices that led to the intervention of the Department of Transportation consisted of deep temporary price cuts directed at specific competitive challenges. In the present instance, the department apparently regards itself as obliged to protect competitors from improvements in service by alliance partners and the possibility that this may result in their loss of business.

—The undeniable benefits to travelers from the merging of frequent flyer programs are indeed highly problematic; and so, indeed, would be the exclusionary consequences of merging override travel agency commissions.

41. Department of Transportation, "DOT Conditionally Allows Delta-Northwest-Continental Code-Sharing Agreement," press release, January 17, 2003. The carriers subsequently accepted this condition. Department of Transportation, *International Aviation Developments*.

42. "First DOT Opinion," pp. 12–13, 15.

43. Ibid., p. 11.

44. For a particularly flagrant illustration of this practice in the conditions attached by the Federal Communications Commission to its approval of "the proposed union of SBC and Ameritech, extract[ing] from the companies thirty 'voluntary' commitments by the parties to *do* certain things,

to *behave* in certain ways—subject to possible liability for 'voluntary incentive payments' . . . of as much as $1.125 billion—some of them only remotely and others not even remotely related to the perceived possible injuries to competition *stemming from the merger*," accompanied by eloquent, devastating dissents by Commissioner Harold Furchtgott-Roth and then-commissioner—now chairman—Powell, see Kahn, *Whom the Gods Would Destroy*, pp. 39–45.

45. Note that American, for example, merely acquired the assets of TWA out of bankruptcy. Conceivably that is the form other consolidations might take.

46. Julius Maldutis, "Statement," before the Senate Committee on Commerce, Science, and Transportation, November 4, 1987, p. 9. To be sure, this did not involve nine separate competitors. Alfred E. Kahn, "Surprises of Airline Deregulation," *American Economic Review,* vol. 778 (May 1988), pp. 316–22: "The industry remains to this very day far more intensely competitive than it was before 1978. The opponents of deregulation cannot have it both ways—asserting, on the one hand, that competition has proved to be a lost cause and, on the other, that it has been and remains catastrophically destructive. They will undoubtedly retort that the process of competition killing itself off is still incomplete. The response—now, as ten years ago—is that the possibility, which no one can deny with total certainty, that competition *may* one day prove not to be viable is hardly a reason to have suppressed it thoroughly in the first place." (p. 319)

47. Coincidentally, and likewise directly relevant to this example, is the requirement by the Department of Transportation—of which I have already approved—that the parties to the Continental-Delta-Northwest alliance free up some gates at Logan Airport, as a condition of its approval. See note 41, above.

48. See the astoundingly high estimates of the costs of regulatory delay in approving wireless services by Jerry A. Hausman, "Valuing the Effect of Regulation on New Services in Telecommunications" (Brookings, 1997); Jeffrey H. Rohlfs, Charles L. Jackson, and Tracey E. Kelley, "Estimate of the Loss to the United States Caused by the FCC's Delay in Licensing Cellular Telecommunications" (White Plains, N.Y.: National Economic Research Associates, November 4, 1991. See also Robert W. Crandall, Robert W. Hahn, and Timothy J. Tardiff, "The Benefits of Broadband and the Effect of Regulation," in Robert E. Crandall and James H. Alleman, eds., *Broad-*

band: Should We Regulate High-Speed Internet Access? (AEI-Brookings Joint Center for Regulatory Studies, 2002).

49. Robert W. Crandall and Leonard Waverman, *Talk Is Cheap* (Brookings, 1996). The full meaning of the title of their book is "Talk *Ought to Be* Cheap but Isn't because of Our Perverse Regulatory Policies." For a pioneering effort to estimate the welfare losses, see reference to L. Perl, "Impacts of Local Measured Service in South Central Bell's Service Area in Kentucky 24," prepared by National Economic Research Associates for South Central Bell Telephone Company, May 21, 1985 (unpublished manuscript on file with NERA), in Alfred E. Kahn and William B. Shew, "Current Issues in Telecommunications Regulation: Pricing," *Yale Journal on Regulation*, vol. 4 (Spring 1987), n. 21, p. 197.

50. As of the end of 2002, there were 136 million subscribers to cellular and other mobile wireless service, as compared with 188 million wire lines. Federal Communications Commission, "Local Telephone Competition: Status as of December 31, 2002," Washington, June 2003. Some 6.5 million of the 129 million wireless subscribers noted in an FCC report six months earlier did not even have wired phone service. James Gattuso, "The FCC's Local Competition Report: Surprise!" *C:SPIN,* Competitive Enterprise Institute, January 31, 2003 (www.cei.org).

51. Dennis K. Berman, "Internet Calls Stir up Static in Phone Fight," *Wall Street Journal*, January 24, 2003, p. B1.

52. Crandall and Waverman, *Talk Is Cheap*. The contribution from intraLATA (local access and transport area) toll and inter- and intrastate switched access services alone, measured by the difference between revenues and their *incremental* costs, amounted to $23.6 billion in 1995; see John Haring and Jeffrey H. Rohlfs, "Economic Perspectives on Access Charge Reform," prepared for BellSouth Telecommunications (Bethesda, Md.: Strategic Policy Research, January 29, 1997), p. 13.

53. In deregulating the airlines, Congress wisely substituted a minimal Essential Air Service program, taxpayer financed, for the previous combination of direct and internal subsidizations. Clearly, it would have been far better had it done the same thing in the case of telecommunications.

54. See the powerful argument of Robert W. Crandall that the exclusion of the Baby Bells from long-distance service was not necessary, citing the experience of Canada and the United Kingdom, which experienced more rapid growth of competition than the United States, while imposing

no such requirements or prohibitions. Crandall, "The Failure of Structural Remedies in Sherman Act Monopolization Cases," *Oregon Law Review*, vol. 80 (Spring 2001), sec. IIIG. Crandall's contention is that the equal access requirement would have been sufficient. On the undoubted essentiality of mandatory interconnection, in contrast, see Eli M. Noam, *Interconnecting the Network of Networks* (MIT Press, 2001). See also note 111, below.

55. See Jerry A. Hausman, Gregory K. Leonard, and J. Gregory Sidak, "Does Bell Company Entry into Long-Distance Telecommunications Benefit Consumers?" *Antitrust Law Journal*, vol. 70, no. 2 (2002), pp. 463–84.

56. Joint Affidavit of Robert Crandall and Leonard Waverman on Behalf of Ameritech Michigan, "In the Matter of Application of Ameritech Michigan Pursuant to Section 271 of the Telecommunications Act of 1996 to Provide In-Region InterLATA Services in Michigan," FCC CC Docket no. 97-1, vol. 3.1, January 2, 1997, paras. 53–54.

57. Frederick Warren-Boultin, "Direct Testimony before the Kansas Corporation Commission on behalf of AT&T and MCI," May 12, 1998, cited in Alfred E. Kahn and Timothy J. Tardiff, "Testimony before the Public Service Commission of the State of Missouri in the Matter of Application of SBC Communications, Inc., Southwestern Bell Telephone Company, and Southwestern Bell Communications Services, Inc., d/b/a Southwestern Bell Long Distance for Provision of In-Region, InterLATA Services in Missouri, on behalf of Southwestern Bell Telephone Company," Docket no. TO 99-227, filed November 20, 1998; Surrebuttal Affidavit, February 1, 1999. See also William Taylor, "Declaration Regarding Competition in Massachusetts and the Public Interest Benefits of InterLATA Entry, FCC, in the Matter of Application by Verizon New England Inc., et al. for Authorization to Provide In-Region, InterLATA Services in Massachusetts," appendix A, September 19, 2000, paras. 9–12.

58. See note 55, above. Also, Hausman, Leonard, and Sidak, "Does Bell Company Entry into Long-Distance Telecommunications Benefit Consumers?" especially pp. 482–84.

59. See Robert W. Crandall, "Debating U.S. Broadband Policy: An Economic Perspective," Brookings Policy Brief 117 (March 2003), p. 7.

60. There have been some heroic exceptions at the state level—cases in which commissions have recognized the awesome waste in overcharging for long-distance service. Two heroes at the federal level were FCC chairman

Mark Fowler and the chief of his Common Carrier Division, Albert Halprin. They recognized in the very early 1980s that the inflated access charges not only were grossly inefficient, but were no longer sustainable in the face of growing competition among access providers. In a courageous move, they induced the FCC to substitute flat subscriber line charges for usage-sensitive access fees. Although Congress forced them to retract part of those flat charges, their initiative constituted a very important step in the direction of greater efficiency. See Mark S. Fowler, Albert Halprin, and James D. Schlichting, "Back to the Future: A Model for Telecommunications," *Federal Communications Law Journal,* vol. 38 (1986), p. 135; Gerald W. Brock, *Telecommunications Policy for the Information Age* (Harvard University Press, 1994). The FCC later adopted a considerably less courageous (and less efficient) method for financing otherwise laudable further reductions in access charges—a move proclaimed by the chairman as responsible for "the single best day for consumers in the agency's history." See the account in Kahn, *Letting Go,* pp. 134, 138–44.

61. "Competition for Special Access Services," Attachment B to "Opposition of Verizon, before the Federal Communications Commission, in the matter of AT&T's Petition for Rulemaking to Reform Regulation of Incumbent Local Exchange Carrier Rates for Interstate Special Access Services," RM 10593, December 2, 2002.

62. See the annual reports of the Association for Local Telecommunications Services (ALTS), "The State of Local Competition" (Washington, 2001–03). Kevin A. Hassett and Laurence J. Kotlikoff show CLECs investing $17.2 billion and $21.7 billion annually in 1999 and 2000, some 43 percent of the industry total; and $11–12 billion in the next two years, 30–31 percent of the industry total. Kevin A. Hassett and Laurence J. Kotlikoff, "The Role of Competition in Stimulating Telecom Investment," October 2002, p. 8 (www.aei.org). Debra Aron has demonstrated (see note 100, below) that the decline in facilities-based lines in the last two years in Illinois was the result not just of the recession but, for obvious reasons, of the almost total shift of the CLECs to using the recently available UNE-P (unbundled element network platforms) that SBC Illinois was required by that state's Commerce Commission to offer them, at blank slate TELRIC (total element long-run incremental cost) prices, in preference to constructing their own facilities. See my discussion of the UNE-P, note 84, below.

63. BellSouth, SBC, Qwest, and Verizon, "UNE Fact Report 2002," submitted "In the Matter of Review of the Section 251 Unbundling Obligations of Incumbent Local Exchange Carriers," CC Docket no. 01-338 (April 2002); also their later supplement, "In the Matter of the Review of the Section 251 Unbundling Obligations of Incumbent Local Exchange Carriers" (October 2002), in the same docket, covering the first six months of 2002.

64. Observe how convoluted the arguments in adversarial regulatory proceedings get: (a) the local Bell companies cited these investments as demonstrating that local competition was proceeding; (b) their adversaries used that same evidence to discredit the contentions of the former that TELRIC pricing discouraged facilities-based competition; (c) the Bells were then constrained to point out that that particular competition was artificially induced by the distorted rate structures.

65. BellSouth and others, "UNE Fact Report," p. I-6. On the additional discouraging effect on facilities-based competitive entry from 2000 onward of the newly available UNE-Platform, at those same TELRIC charges, see below.

66. Sidak, "The Failure of Good Intentions."

67. In analyzing the terrible fatality rate among the approximately 300 facilities-based CLECs in operation at the end of 1999 and the 92 percent decline in their combined market value by the end of 2001, Darby, Eisenach, and Kraemer (*The CLEC Experiment*, p. 4), conclude "that the failure of the CLECs has not slowed down the development of competition for local telephone services," including the remarkable growth of cable telephony.

68. An outstanding practitioner of dishonest accounting was MCI WorldCom, which evidently reported tens of billions of dollars of fraudulent investment by the simple expedient of capitalizing expenses. In March 2003 the company announced it was "writing down $79.8 billion of good will and other assets." See Simon Romero, "WorldCom Decides to Take $79 Billion Write-Down," *New York Times*, March 14, 2003, p. C2.

69. See notes 49 and 53, above; Alfred E. Kahn, *The Issue of Cream Skimming*, vol. 2 of *The Economics of Regulation*, reprint (MIT Press, 1988 [1970, 1971]), pp. 220–46; Alfred E. Kahn, "The Road to More Intelligent Telephone Pricing," *Yale Journal on Regulation*, vol. 1, no. 2 (1984), pp. 139–57; and for an analysis of the proffered "economic" rationale of the

previous rate structures, Kahn and Shew, "Current Issues in Telecommunications Regulation," pp. 191–256.

70. Among which, I understand, have been the incumbent local and long-distance carriers, with whom the CLECs ran up huge bills for wholesale purchases on which they then defaulted. I mean in no way to minimize the insecurity suffered also by employees.

71. John Wohlstetter cited an estimate by Merrill Lynch in 2001 that only 2.5 percent of fiber capacity was being used. John Wohlstetter, "Fiber Fables II: The Long Distance Fiber Glut Is Last-Mile Copper Scarcity," Discovery Institute, November 5, 2001. He has also called to my attention an estimate of 5 percent in April 2003.

72. U.S. Supreme Court, *AT&T Corporation et al.* v. *Iowa Utilities Board et al.*, 525 US 366 (January 25, 1999); U.S. Supreme Court, *Verizon Communications, Inc. et al.* v. *Federal Communications Commission et al.*, 122 S. Ct. 1646, certiorari to the U.S. Court of Appeals for the Eighth Circuit, May 13, 2002.

73. Alfred E. Kahn, *The Economics of Regulation*, 2 vols. (John Wiley, 1970, 1971; reprint, MIT Press, 1988).

74. Hausman, "Valuing the Effect of Regulation on New Services in Telecommunications."

75. My argument here followed the earlier classic exposition by William Fellner, "The Influence of Market Structure on Technological Progress," in American Economic Association, *Readings in Industrial Organization and Public Policy* (Homewood, Ill.: Richard E. Irwin, 1958), pp. 287–91.

76. Alfred E. Kahn, Timothy J. Tardiff, and Dennis L. Weisman, "The Telecommunications Act at Three Years: An Economic Assessment of Its Implementation by the Federal Communications Commission," *Journal of Information Economics and Policy*, vol. 11 (1999), p. 337; and Timothy J. Tardiff, "Cost Standards for Efficient Competition," in Michael A. Drew, ed., *Expanding Competition in Regulated Industries* (Boston: Kluwer, 2000), p. 180. In a recent informal survey, Tardiff discovered that the *monthly* rental price of a personal computer is on the order of one-sixth to one-third of its purchase price. The implied payback period of three to six months reflects the fact that computer prices are declining rapidly (the producer price index for personal computers has declined at an annual rate of 30 to

40 percent), as well as the operating costs of a computer rental firm and uncertainty in the demand for its services.

77. See note 72, above. Gregory L. Rosston and Roger C. Noll, "The Economics of the Supreme Court's Decision on Forward-Looking Costs," *Review of Network Economics*, vol. 1 (September 2002), pp. 81–89.

78. "The Commission has in effect declared: '*We* will determine not what your costs are or will be but what we think they *ought to be*. Why should we bother to let the messy and uncertain competitive process determine the outcome when *we* can determine at the very outset what those results would be and prescribe them now?'" (Kahn, *Letting Go*, p. 92).

79. It is not merely infatuation with my own metaphors that impels me to repeat my previous observations to the effect that the FCC likewise succumbed to the temptation to use its oversight of merger proposals to impose detailed conditions for approval (see notes 38 and 44, above), some clearly inspired by Robin Hood, enforced by elaborate systems of fines for noncompliance: SBC paid over $60 million in penalties for allegedly violating the conditions of its merger with Ameritech. One example of the commission's egregious micromanagement was inclusion of its pet formula for conducting broadband operations under the aegis of fully supported subsidiaries. Kahn, *Whom the Gods Would Destroy*, pp. 40–45.

80. "Stranded Costs and Regulatory Opportunism," in ibid., pp. 31–34; "Where the Money Is, Temptations of the Kleptocrats," in Kahn, *Letting Go*, pp. 9–15.

81. "The average X-factor (reflecting achievable annual improvements in productivity) adopted by . . . state commissions . . . was 2.5 percent. At this average annual rate of cost reduction, it would take at least 28 years to achieve the cost reductions embodied in the UNE rates set by the states. . . . Yet I am not aware of a single finding in a state UNE proceeding that the incumbent has operated inefficiently." Dale E. Lehman, "The Court's Divide," *Review of Network Economics*, vol. 1, no. 2 (2002), pp. 108–09. See the similar claim of the Canadian TELUS Communications in protesting the charges to CLECs for use of its unbundled loops, prescribed by the Canadian Radio-Television and Telecommunications Commission, claiming that it involved an immediate 20–25 percent reduction, while also continuing from that point onward the further 3.5 percent real decrease incorporated in its previously imposed caps. Here, evidently, was a similar seemingly egregious double counting of what is supposed to be achievable.

Alfred E. Kahn, "Petition of TELUS Communications, Inc., to the Governor in Council, Government of Canada, to Vary Telecom Decision," CRTC 2002-67, January 27, 2003, Appendix E.

82. This statement should elicit a good laugh from anyone familiar with how delivery on the stranded cost promise was frustrated in the cases of Pacific Gas and Electric and Southern California Edison by the collision between skyrocketing wholesale electricity costs and retail price ceilings. Note, though, that San Diego Gas and Electric was able to recover stranded costs before the deluge.

83. In (grudging) fairness, I must point out that Congress provided sufficient justification by instructing the commission to set rates for the UNEs based on return on investment, "without reference to a rate of return or other rate-based proceeding." It would be interesting to trace the legislative history, to ascertain at whose insistence this open invitation to expropriation was inserted. On the other hand, the act authorized the state commissions to add a "reasonable profit," which obviously left room for the exercise of discretion in defining that term. Moreover, the widespread substitution of direct price cap regulation for rate-based rate of return regulation over the preceding several years was rationalized as a means of producing utility rates that approximated economically efficient levels. And while the initial caps were, as I understand it, typically based on rates found "just and reasonable" in traditional proceedings, it was clearly expected that within a few years rates would be free of the taint of traditional cost-plus rate base/rate of return. I suggest also that the Supreme Court's decision in 2002 sustaining TELRIC offers sufficient basis to believe that the FCC had the legal discretion, had it so chosen, to set rates on the basis of the companies' *own* long-run incremental costs, as the earlier Circuit Court of Appeals decision had required. *Iowa Utilities Board et al.* v. *Federal Communications Commission*, 219 F 3d 744 (2000).

84. Verizon has contended that the requirement on ILECs to make their products available to their competitors at wholesale prices—whether under simple resale agreements or by using the UNE-platform—had the effect not merely of permitting those independents to compete with it, but of *excluding* Verizon from the retail market: "AT&T was a reseller of service physically being furnished by Verizon regardless of which of two options it chose under the agreement: purchasing Verizon 'service' at a wholesale price for 'resale'; or purchasing the end-to-end package of Verizon-assembled

piece parts (UNEs) commonly called 'UNE platform.' . . . Whereas 'service' is bought in the former arrangement, the latter involves local loop *occupation* by the competitor, because purchase of a loop as an 'unbundled network element' gives the competitor *exclusive* occupation of that piece of the incumbent's network." *Verizon Communications Inc.* v. *Law Offices of Curtis V. Trinko*, 305 F3d 89 (2d Circuit Court of Appeals, 2002), and Petition for Writ of Certiorari, U.S. Supreme Court (for Verizon), November 1, 2002, p. 4, note 4.

This protest betrays an intuitive notion that application of the (something like) essential facilities doctrine to subscriber access lines was in some ways fundamentally different from its application in all other instances with which I am familiar—and in ways almost certain to generate resistance and conflict. Requiring monopolists to make their facilities, goods, or services available to competitors does not in itself prevent the owners from competing in the market. Compulsory licensing of patents does not in itself exclude the patent owner from using the patents and selling the products or services in competition with its licensees. A similar argument can be made with respect to the requirement that railroads give competitive access to bottleneck trackage; or the requirement that Aspen Skiing include the Highlands facilities in its all-hills one-week passes; or that ILECs interconnect with competitors. By contrast, in the loops case, owners are required to lease their lines to competitors on an *exclusive* basis: if the CLEC wins over the customer, it effectively takes over exclusive right to the access lines, and with it the opportunity to offer subscribers such typically overpriced services as call waiting and caller ID. Superficial appearances to the contrary, "line sharing" would not entail an exception to or a contradiction of this observation. True, line sharing contemplates that the ILEC and the CLEC would share the "line." But the CLEC receiving sharing rights obtains the *exclusive* right to the high-frequency capabilities of the line, leaving to the ILEC—or a CLEC lessee—the continued exclusive use of the capabilities of the line for voice-grade service for that particular customer.

In fact, I see nothing in the exclusive control of the subscriber access line by incumbents or by competing lessees that excludes the possibility of competition between them. It merely reflects the fact that competition in the local telephone business is competition for the *patronage* of the subscriber, whether or not employing the access lines owned by the ILECs. The competition takes the form of alternative packages of services, with the winners, necessarily, having exclusive use of the subscriber access lines—or, in the case of line sharing, the particular high- or low-frequency capabilities of those lines. So long as that contract is rescindable—that is to say, so long as the incumbent company is free to try to lure back former subscribers by offering them more attractive packages of services—there would seem to be no greater legitimacy to the resistance of incumbents to "line sharing" than to mandatory leasing of the customer access lines, if there is to be efficient competition at the local level. The ILECs have apparently not objected in principle to the requirement of the Telecommunications Act that they stand ready to lease their lines to would-be competitors—even though the lease then excludes them from use of the lines to reach the departing customers. Therefore, their legitimate objections to the FCC's rulings with respect to access lines are directed properly at the absurd cost-determination methods it has prescribed.

85. See text associated with note 63, above. The FCC's adoption of its TELRIC blank-slate pricing standard was additionally ironic because it was *unnecessary* to encourage competition. As witnesses for AT&T, a leading proponent of TELRIC pricing, have pointed out time and again in other contexts, what matters for efficient competition is not the absolute level of the charges for the incumbent's inputs, but the requirement that those charges also be incorporated in the prices of the incumbents' own services—that is, that the *margin* between wholesale charges by incumbents to would-be retail competitors and their own retail prices comply with what the AT&T witnesses termed the "efficient component pricing principle," reflecting the full long-run marginal costs of the incumbents. Doubly ironic, in AT&T's original enunciation of that pricing principle in New Zealand, these same witnesses contended that the incumbent, New Zealand Telecom, was entitled to charge would-be competitors its full "opportunity costs"— the contribution to common costs and profits lost by selling to them at wholesale rather than to itself at retail. That argument was then eagerly adopted by some of the ILECs in Section 251 proceedings, which proposed that the prices of the unbundled network element inputs be set in accordance with AT&T's witnesses' New Zealand version of the "efficient component pricing principle." The ILECs found to their dismay that AT&T's witnesses now argued that the economically proper basis for pricing was TELRIC. By contrast, in those original New Zealand proceedings, I argued for a strict separation of the issue of competitive parity (which required only

that whatever markup above their incremental cost was incorporated in the access charges also be incorporated in the incumbents' retail prices, along with its own incremental costs of performing the retail function), and that the determination of the proper *level* of the markup was a *regulatory* matter, to be determined by some other authority, on the basis of whatever understandings about cost recovery there may have been in New Zealand at the time of privatization. See Alfred E. Kahn and William E. Taylor, "The Pricing of Inputs Sold to Competitors: A Comment," *Yale Journal on Regulation*, vol. 11 (1994), pp. 226–40; and the gracious acknowledgment of the point by William Baumol and J. Gregory Sidak in their "Response," *Yale Journal on Regulation*, vol. 12 (1995), pp. 177–86.

86. The FCC's report on local competition June 12, 2003 (tables 3 and 5) provides empirical support for this proposition: between December 2000 and December 2002, the number of CLEC-owned access lines increased from 5.2 million to 6.4 million. But cable telephony subscribers increased from 1.1 million to 3.0 million, implying that the number of non-cable access lines fell by some 700,000. This should not be surprising, since this was the period in which the use of UNE-Ps increased from 2.8 million to 10.2 million. This is one more indication that the commission's eagerness to promote "competitors" with its TELRIC pricing had the predictable effect of discouraging the facilities-based competition that it had itself proclaimed so important. Yet another indication is provided by the FCC's June 12, 2003, report on local telephone competition: between 2001 and 2002, the share of CLEC-operated lines that it defines as facilities-based declined from 50 to 45 percent of the total. See tables 3 and 4.

Looking specifically at developments in Illinois, Aron discovered, unsurprisingly, that once UNE-Ps became available, additional recourse to facilities-based CLEC competition "plummeted." See note 100, below, and associated text.

See the fuller exposition and documentation of this negative effect of the UNE-P and its consequent negative effect on growth in the economy at large in Stephen B. Pociask, "The Effects of Bargain Wholesale Prices on Local Telephone Competition: Does Helping Competitors Help Consumers?" (Washington: Competitive Enterprise Institute and New Millennium Research Council, June 2003).

Significantly, when a story in the *Wall Street Journal* predicted that the FCC was about to phase out the UNE-P, the prices of facilities-based

CLECs' stocks actually increased somewhat more than those of the Bell companies.

87. The compelling criticisms of FCC policy and comprehensive proposals for its reform by John C. Wohlstetter would seem at first blush to have missed the commission's entitlement to claim that it was right, albeit for the wrong reasons: "The so-called Competitive Local Exchange Carriers (CLECs) imploded because the vast majority of the more than two hundred firms that entered the marketplace after passage of the 1996 Telecom Act offered services duplicating those already available from existing phone companies, hoping to take advantage of the artificially low prices mandated by the Clinton FCC. Lacking better technology and without real ideas for new services, the firms collapsed." John C. Wohlstetter, "Telecom Meltdown," *American Outlook* (Fall 2002), p. 3.

The UNE-P obviously gave CLECs an additional resale option not explicitly provided in the act, an alternative of particular value to them in competing to sell services that were overpriced by regulatory dictate. For underpriced services, they retained and have exercised the resale option explicitly provided in the act—to purchase at discounts from those low regulated retail rates at generous discounts prescribed by the FCC. See Kahn, *Letting Go*, pp. 96–98.

88. See the description of even more intensified pursuit of the latter policies by several states in 2002 at notes 99–105, below.

89. See notes 62 and 87, above.

90. That sequence, spelled out more clearly by Hazlett, "The Irony of Regulated Competition in Telecommunications," pp. 11–17, appears to be also what Wohlstetter had in mind (see note 71, above). While congenial to my argument, this reconciliation seems to ignore the fact that the bulk of the facilities-based CLECs concentrated on services to businesses, especially large businesses, in metropolitan areas, whereas the underpriced UNE-Ps were aimed at encouraging competition in bundled services to residential customers.

91. "Separate Statement of Chairman Michael K. Powell, Dissenting in Part," attached to FCC, "FCC Adopts New Rules for Network Unbundling Obligations of Incumbent Local Phone Carriers," press release, February 20, 2003 (emphasis added). Recall that only a minority of the UNEs used by CLECs serving residential customers included their own switches. According to the 2002 "UNE Fact Report," there were 8.6 million residential lines

served by CLECs at the end of 2001. Of these, 5.6 million were UNE-P or
resale: that is, they were served with an ILEC switch. The other 3 million
were "facilities-based," which according to the "UNE Fact Report" definition means that they used at least a CLEC switch—in some (but not all)
cases, CLEC loop facilities were used as well.

92. "Statement of 43 Economists on the Proper Regulatory Treatment
of Broadband Internet Access Services," before the Federal Communications
Commission, in the matters of "Review of Section 251 Unbundling Obligations of Incumbent Local Exchange Carriers," CC Docket no. 01-339;
"Review of Regulatory Requirements for LEC Broadband Telecommunications Services," CC Docket no. 01-337, and "Appropriate Framework for
Broadband Access to Internet over Wireline Facilities," CC Docket no.
02-33, May 3, 2002. For a skeptical view of the extent to which the sharing
obligation has been responsible for retarding investment in broadband or
that actual deployment has been short of reasonable expectations, see Gerald R. Faulhaber, "Broadband Deployment: Is Policy in the Way?" in Crandall and Alleman, *Broadband,* pp. 223–44. Faulhaber does agree, however,
that "applying legacy regulation to new technologies is almost surely inap-
propriate." In this he agrees with a recent report of the National Academy of
Sciences that while "it is reasonable to maintain unbundling rules for the
present copper plant. . . . Existing bundling rules should be relaxed only
where the incumbent makes significant investments . . . to facilities constructed to enable new capabilities" (quoted on p. 243).

93. John Wohlstetter graphically points out the relationship of this
reform to the glut of *long-distance* fiber capacity: "bandwidth is the instan-
taneous throughput of an end-to-end connection. Just as a chain is no
stronger than its weakest link, a network connection is no broader than the
narrowest segment of an end-to-end link." (Wohlstetter, "Fiber Fables II.")
And that is probably the most suitable final word in this section attempting
to assess the role of the Section 251 process in achieving the goals of the
Telecommunications Act. I am indebted to Thomas Hazlett for making it
clear to me: *not one of the unequivocal successes of deregulation that I summa-
rized at the very outset of this section depended on the sharing of UNEs prescribed by the FCC.* Hazlett, "The Irony of Regulated Competition in
Telecommunications," pp. 2, 5.

94. But see the skeptical view of Faulhaber in note 92, above.

95. *AT&T Corp.* v. *Iowa Utilities Board,* 525 US 366 (1999); *U.S. Telecom Assn.* v. *FCC et al.,* 290 F3d 415 (U.S. Court of Appeals for the District of Columbia Circuit, May 24, 2002).

96. "The truth is that in the course of our deliberations we never addressed the merits of whether a competitor was actually impaired without access to the switching element and therefore should be unbundled. Instead, the focus of the majority was merely on giving the states a subjective and unrestricted role in determining the fate of the switching element, and therefore UNE-P." (Michael K. Powell, FCC chairman, oral statement before the Telecommunications and Internet Subcommittee, U.S. House of Representatives, February 26, 2003.)

97. See the evocative prediction of "a long stay in regulatory purgatory" by Robert J. Samuelson, "Telecom Purgatory," *Washington Post,* March 6, 2003, p. A23.

As Commissioner Kathleen Q. Abernathy put it in the press statement (February 20, 2003) explaining her dissent from this part of the FCC's 3-2 decision: "While reasonable minds can differ about the appropriate conclusions to draw from the record, and line-drawing is undoubtedly difficult, the Commission was bound to make *some* effort to analyze the data on switch deployment and alleged impairments. For example, the Commission could have made impairment findings based on wire center density. . . . We alternatively could have focused on a threshold number of switches deployed in a LATA or wire center. . . . Another approach would have made UNE-P available as an acquisition tool to give competitors a limited period to aggregate a base of customers before transitioning to UNE-L. . . . The one thing I was *not* willing to do—which unfortunately is what the majority has done here—was to shirk our statutory obligation to decide the circumstances in which unbundled switching will be available. . . . The majority's decision to refrain from adopting a concrete standard for unbundled switching is the exact opposite of what the telecom economy needs. By prolonging the uncertainty indefinitely, I fear that this Order will deal a serious blow to our effort to restore rational investment incentives. While the President and Congress are striving to provide an economic stimulus, the majority unfortunately has stymied that effort. . . . While lawyers will thrive in this environment, the carriers will become mired in a regulatory wasteland. . . . Rather than developing sound business plans in response to the Commission's decision, carriers will spend the next several years in litigation before the state commissions and in the federal district courts.

"In addition to jettisoning the principle of regulatory certainty, the majority's decision tramples on the goal of promoting facilities-based competition. . . . The majority instead has established a regime under which UNE-P may remain permanently available in all markets."

98. There was initially some uncertainty whether, in rushing back in 1996–97 to formulate its rules for the identification and pricing of UNEs, the FCC had not short-circuited the process contemplated by the law: the statute seemed to contemplate a process beginning with negotiation among the affected parties, with resolutions and rules adopted first by the states subject to eventual FCC approval or disapproval. Although the Supreme Court supported the FCC's actions as a reasonable exercise of that ultimate authority, Raymond Gifford, then chairman of the Colorado Public Service Commission, has eloquently suggested how the outcome in that case might have differed: "The Illinois Commission has stopped SBC's DSL rollout dead in its tracks by extending the unbundling requirement to this ridiculous extent. And yes it is terrible policy. This is where I add at least one cheer for federalism: Had not the FCC prescribed both the extent and pricing method and allowed the states some latitude on this, then we'd be collectively better off than the big, national mistakes that Reed Hundt made. For instance, some states would go with TELRIC, others with ECPR; some states would hold close to Professor Areeda's definition of essential facilities for 'necessary and impair,' other states would unbundle everything in the network. Had this trial-and-error federalism been indulged, then the folly of some states would gradually give way to the superior policy of others (led, of course, by Colorado).

"I must admit that there is an element of 'whose ox is being gored' to this defense of federalism. State commissions, as you know, are not exactly hotbeds of innovation, economic sensibility or, sadly, even minimal competence . . . I am less and less patient with the molasses-like incrementalism with which we approach our task. This is a technologically dynamic market where the very market definitions are eroding before our eyes—deregulate it; make the private parties negotiate the prices and reaffirm that contract and property law notions—not administrative regulation—will govern the relationships going forward." (Private communication.)

My present view, heavily influenced by my substantive disagreements with the rules the commission actually formulated, is that Gifford was right. I agree that is what the FCC should have done in 1996–97. But this view runs counter to the fact that, for the most part, telecommunications is clearly an industry that is nationwide in scope and that needs uniform nationwide rules. The case for a local regulatory role for the states has been that the identification of network elements that have to be unbundled for local competition to be feasible depends upon local circumstances. This lends some plausibility to the FCC's decision to let the states decide. But for the commission to do so now, six years later, is an inexcusable, politically motivated cop-out, opening the door to endless litigation in either fifty states and the District of Columbia, or twelve federal jurisdictions, or both.

99. "Bell Companies Lose Customers to AT&T, MCI," *Wall Street Journal*, December 12, 2002, pp. B1, B4. The article reports that SBC was losing nearly one million local customers a quarter to this "competition" and that the share of the competitors of the local market in Michigan had jumped to more than 20 percent from 4 percent in 1999. It cites independent estimates that SBC and Bell South will lose more than 20 percent of their retail lines during the next two years. AT&T and WorldCom defend those sharply reduced wholesale rates on grounds of reciprocity: "Officials at AT&T and WorldCom counter that they provide a parallel service to the Bells by leasing them portions of their networks, which helps enable the Bells to offer long-distance service" (ibid.). But, of course, that analogy obscures the critical difference—namely, that the latter transactions, reflecting the enormous excess capacity in long-distance transmission and therefore, in a sense, a form of destructive competition among the long-distance companies, were nevertheless freely negotiated between the parties (who presumably were recovering the low incremental costs of activating it), and not imposed by opportunistic regulators.

100. Debra J. Aron, Direct Testimony on Behalf of SBC Illinois, before the Illinois Commerce Commission, Docket no. 02-0864, December 23, 2002, p. 10. See the confirmation of similar findings elsewhere, to the effect that the stipulated UNE-P discounts far exceeded the avoided costs of the ILEC, in Pociask, "The Effects of Bargain Wholesale Prices on Local Telephone Competition," pp. 8–14.

101. I testified pro bono, on my insistence.

102. See the similar finding by Dale Lehman, note 81, above.

103. So, Illinois Public Act 93-0005, enacted in May 2003 (within three days of the public hearing at which I testified, and set aside on June 9, 2003, by the U.S. District Court for the Northern District of Illinois, Eastern Division, Case 03 C3290), reads: "The General Assembly finds that *existing actual* total usage of the elements that affected incumbent local exchange carriers are required to provide to competing local exchange carriers . . . is the most reasonable projection of actual total usage. The Commission, therefore, shall employ current actual fill factors . . . in establishing cost based rates for such unbundled network elements." (p. 2, emphasis added.)

104. The statute therefore includes the provision: "The General Assembly further directs that the Commission shall employ depreciation rates that are forward-looking and based on economic lives as reflected in the incumbent local exchange carriers' books of accounts as reported to the investment community under the regulations of the Securities and Exchange Commission. Use of an accelerated depreciation mechanism shall be required in all cases."

I was informed that AT&T had responded to my testimony by producing a new TELRIC calculation purporting to demonstrate that the Illinois commission–ordered rate reductions were more than fully justified—precisely confirming my prediction several years ago: "It takes little imagination or knowledge of regulatory history to envision the combat-by-engineering-and-econometric-models that the TSLRIC-BS standard has invited. . . . No one with any awareness of the history of the tortured efforts by utility commissions in the early decades of this century to comply with the Supreme Court's instruction that utility company rate bases take into account 'the present as compared with the original cost of construction' could possibly have opened the door to similar exercises of 'calculating' any BS [blank slate] version of incremental cost." (Kahn, *Letting Go*, p. 93 and note 135.) This criticism was expatiated on by Supreme Court Justice Stephen A. Breyer, in *AT&T Corp.* v. *Iowa Utilities Board*, 525 U.S. 366 (1999).

One of the most important achievements in public utility regulation in the first half of the twentieth century was the development and application of uniform systems of accounting. Because of it, the factual assertions upon which I based my recommendations to the Illinois legislature could easily be substantiated or disproved. The rebuttal "I have a model that says otherwise" invites the trip to economists' and econometricians' heaven that Justice Breyer and I had predicted.

Subsequently, on May 30, in support of their appeal against the Illinois legislation to the U.S. District Court for the Northern District of Illinois, Eastern Division (see note 103, above), AT&T, MCI, the Association for Local Telephone Services, and an organization called Voices for Choices, submitted a "Reply Declaration of Janusz A. Ordover and Robert D. Willig" to a declaration filed by me on behalf of SBC in the same proceeding, along the same lines as my direct statement to the legislature. One of their central dismissive arguments was that "Dr. Kahn has been a consistent critic both of the economic theory underlying the FCC's pricing rules and of the FCC's pricing rules themselves" (para. 13).

My criticisms—as the foregoing exposition has demonstrated—in concept amounted to restating the irrefutable proposition that setting rates at the estimated minimum cost of an efficient new entrant would be economically incorrect *unless* accompanied by gross rates of return reflecting the "anticipatory retardation" that investors in competitive markets would practice, in industries characterized by rapid technological progress. As I put it in the small volume of mine that they cite in support of their assertions: "[T]he advocacy of TSLRIC-BS is based on the assumption that this is the level to which effective competition would drive prices. That view is mistaken. . . . [An]other way[s] of putting this proposition would be that . . . firms would incur the heavy sunk costs of investing in totally new facilities, embodying the most recent technology from the ground up, only if prevailing market prices were high enough to provide rapid depreciation of those costs and rates of return that Jerry Hausman has estimated would have to be two to three times current costs of capital." (Kahn, *Letting Go*, pp. 91–92).

The issue reduces, then, to whether the FCC or the Illinois Commerce Commission has by its actions paid adequate respect to the last condition, the legitimacy of which both the federal agency and Supreme Court explicitly recognized.

Moreover, as the subtitle of *Letting Go* clearly suggests, the promulgation of TELRIC poses an obvious "temptation of the kleptocrats," and its application, a clear example of "the political economy of regulatory disingenuousness." My discussion elaborated my conviction that this hypothetical standard presented an irresistible temptation of regulatory agencies to opportunistic behavior, in an effort to demonstrate quick results in the form of rate reductions and live competitors, however subsidized.

The burden of my demonstration and that of Debra Aron in Illinois was that the Illinois Commerce Commission *could not more conclusively have vindicated my predictions and criticisms of the TELRIC standard had it set out deliberately to do so*. Given this record, the gods must surely have been mad— or indifferent—if they had not responded to the argument by Ordover and Willig, apparently seriously offered, with gales of celestial laughter—as the title of my 2001 *Whom the Gods Would Destroy* suggested they would.

105. According to Banc of America Securities, the increase in SBC Illinois' UNE-P charges would cut profit margins from 60 to 40 percent: "A new Illinois law signed by the Governor yesterday has a negligible financial impact on AT&T." David W. Barden and others, "AT&T Corporation: Local Launches in More States; Putting UNE in Perspective," *Research Brief* (Banc of America Securities, May 13, 2003). See also studies cited by Pociask to the effect that UNE-P prices generally recover only 50 percent of the ILECs' retail revenues (and the public estimate of 55 percent by AT&T itself). Pociask, "The Effects of Bargain Wholesale Prices on Local Telephone Competition," p. 8.

106. I can express only chagrin at the apparent disagreement of some academic economists with that proposition. One, representing AT&T, proclaimed that "a move by the Federal Communications Commission to make the rates, terms or availability [of the current network pricing scheme] less attractive at this point will not create more competition but will, rather, almost certainly—and perversely—spell the end of the development of local exchange competition." "A Dream Nears Reality: The Ease-up at the F.C.C.," *New York Times*, Business Section, February 2, 2003, p. 7.

There is a similar consistent strain in Robert E. Litan's "The Telecom Crash: What to Do Now?" *Policy Brief* 112 (Brookings, December 2002). After describing the drastic reductions in prescribed UNE-P prices in the several states to which I have already referred, he characterizes the objections of the regional Bell operating companies as a continuing effort "to limit the competition," to "curtail" their "current unbundling obligations." He simply passes over the fact that it is not the unbundling to which the regional Bell operating companies object (at least, overtly)—it is the *complete bundling* involved in the oxymoronic "unbundled network elements (*plat-*

form)." "Many providers argue that their financial interests should be the primary objective," Litan asserts—ascribing to them unbelievable ineptness in public relations (p. 4). And he offers the remarkable characterization of the policy issue and his implied prescription: "[W]hat, if anything, should policymakers do now? Should they adopt policies explicitly aimed at assisting the telecommunications providers [that is, the ILECs], or should they put the consumers' interests first? . . . [T]he purpose[s] of having firms in a capitalist economy [is] *to serve consumers*, which is best done by allowing *natural* competitive forces to work so that the fittest ones that deliver the best service at the lowest cost survive.

"If state regulators are allowed to implement pro-consumer *unbundling* [!—recall, the ILECs' critical objections are to the UNE-P] requirements . . . even local telephone service rates may fall." (p. 8, emphasis added.)

The resemblance between the substantive demands of the ILECs and the foregoing characterization of them, and between the recent drastic reductions in *regulated* UNE-P rates and "natural competitive forces" is difficult to discern. Ironically, in the same article, Litan graphically describes the entirely beneficent and powerful competitive effect of the growth of wireless telephony—unsubsidized by ILECs and subject only to the requirement of reciprocal interconnection.

107. Alfred E. Kahn, "Bribing Customers to Leave and Calling It 'Competition,'" *Electricity Journal* (May 1999), pp. 88–90.

108. After noting the lead taken by commissions in California, New York, Michigan, and other states in drastically reducing UNE rates, Hassett and Kotlikoff ("The Role of Competition in Stimulating Industry Investment," p. 10) quote one recent estimate "that local phone customers that switch to CLEC providers can save $11.40 a month in California," citing this and other such cases of achievable savings of many hundreds of dollars each year as concrete evidence of the benefits of "competition"! See also Stephen Labaton, "Dream Nears Reality: Ease Up at the F.C.C.," *New York Times,* February 2, 2003, sec. 2, p. 1.

109. "The Great Telecoms Crash," *Economist,* July 20, 2002, p. 9.

110. Simon Romero, "The Baby Bells Shake the Industry as They Lose Ground to Local Phone Companies," *New York Times,* October 7, 2002, p. C2. The subsequent 20–25 percent recovery left their decline still in excess of 25 percent.

111. I confess to uncertainty about the validity of the analogy between CLECs and wireless carriers. To the extent that all wireless service providers require is mere interconnection with the ILECs for mutual initiation and termination of calls—which Crandall contends would have been sufficient (see note 54, above), without divestiture, for the success of competition in long-distance service as well—the opportunities for friction between the ILECs and their wireless competitors were clearly less than the friction associated with the mandatory provision of UNEs to the CLECs. On the other hand, wireless carriers also buy large amounts of special access services from the ILECs, which they have been able to negotiate (under FCC mandate but not at TELRIC prices), with excellent success and without apparent friction.

112. *Verizon Communications Inc.* v. *Law Offices of Curtis V. Trinko*, 305 F3d 89 (2d Circuit Court of Appeals, 2002); see note 84, above.

113. See, for example, Alfred E. Kahn, "The Uneasy Marriage of Regulation and Competition," *Telematics* (September 1984), pp. 1–2, 8–17. [Editor's note: This note is numbered 66 in the source document.]

114. Kahn, *Whom the Gods Would Destroy*, pp. 35–36.

115. Ibid., p. 36.

116. E-mail addressed to the Advisory Board of the American Antitrust Institute on or about March 10, 2003. The 1996 act did not confine the obligation of ILECs to share UNEs to carriers that had previously been found to have violated the antitrust laws, or to essential facilities, as understood under the antitrust laws. Rather, it entrusted to the FCC responsibility for determining which facilities were "necessary" for competitors, or which the lack of would "impair" their ability to compete, and imposed its own remedies and sanctions, without regard to antitrust culpability.

117. See Kahn, *Whom the Gods Would Destroy*, p. 23. "Mr. Powell's view actually was shared by four of the five FCC commissioners, who said they would have preferred to keep line-sharing regulations intact. But the commissioners agreed to give up line-sharing in exchange for keeping local voice-calling competition in place as part of the complex, highly contested ruling." Julia Angwin, "FCC's Ruling Could Deal Blow to Rural ISPs," *Wall Street Journal,* February 25, 2003, p. B1. That the two dissenters really reflected the majority view on this issue is further confirmed by the press statement, "Approving in Part, Concurring in Part, Dissenting in Part," of Commissioner Michael J. Copps: "There are aspects of this Order that are

certainly not my preferred approach, but which I have had to accept in order to reach compromise. In particular, there is the decision to eliminate access to only part of the frequencies of the loop as a network element. I would have preferred to maintain this access, also known as line sharing. I believe that line sharing has made a contribution to the competitive landscape." (*FCC Headlines 2003*, February 20, 2003, p. 2; www.fcc.gov.) A peculiar, but revealing, confession.

118. See Verizon's complaint in its petition for certiorari in note 84, above.

119. "The FCC Presses Auto-Destruct," *Economist*, March 1, 2003, p. 56; and Angwin, "FCC's Ruling Could Deal Blow to Rural ISPs."

120. On the other hand the cable companies have offered the persuasive defense of their exemption that reconfiguring their cable infrastructure for two-way telephony has been costly and risky and would be discouraged by mandatory sharing. See Hazlett, "The Irony of Regulated Competition in Telecommunications," p. 5.

Charles A. Zielinski reminds me of another asymmetrical regulatory policy that similarly distorts the competition between the local telephone and cable companies in the offer of high-speed Internet access to the disadvantage of the former: the FCC's policy of charging cable companies less than telephone companies for attachments to the former's poles or conduits—in direct violation of the principle that I enunciated in my *Economics of Regulation* (vol. 1, pp. 174–75), drawing in turn upon James C. Bonbright's classic *Principles of Public Utility Rates* (Columbia University Press, 1961). With no apparent reconsideration, the FCC reaffirmed this policy in its implementation of the expanded pole attachment rate-setting authority granted to it by the 1996 act. The Supreme Court's opinion sustaining the authority of the FCC to set rates charged the cable companies for the carriage of both traditional cable service and high-speed Internet traffic did not address the merits of the fees stipulated or the differences between them. *National Cable & Telecommunications Association, Inc.* v. *Gulf Power Co., et al.*, 534 US 327 (2002). As I have pointed out elsewhere, the difference in the rates may have made sense in terms of the respective elasticities of demand for telephone and traditional cable service, but not for high-speed access provided by the two categories of carriers in competition with one another. See Kahn, *Whom the Gods Would Destroy*, p. 13, note 27.

121. See Almar LaTour and Peter Grant, "Verizon May Set Off Price War, Decision to Cut DSL Rates May Put Pressure on Cable Firms," *Wall Street Journal,* May 5, 2003, p. B2.

122. See Marcello Prince, "Dialing for Dollars," *Wall Street Journal,* May 18, 2003, p. R8.

Index

Abernathy, Kathleen Q., 33, 34, 43, 73n97

Air Carriers Association of America, 13

Airline industry: code-sharing alliances, 7–18, 52n13, 53n17, 54n18, 55n24, 57n30, 58n39; costs and losses, 4, 5–6, 51n10; cyclical sensitivities, 4; economic and competitive factors, 9–11, 18–19, 53n17, 56n27; effects of *9/11*, 4–5, 7, 8; exclusionary power, 10–11; fares and pricing, 10, 13–14, 15, 52n14, 55n24, 57n30, 57n33, 59n40; foreign ownership, 5; government assistance to, 4–5; interlining, 8, 10; low-fare carriers, 6, 10, 16, 18; market capitalization, 49n1; mergers, 17–19; open-skies agreement, 53n17; point-to-point flights, 19; restructuring responses, 7–19; security factors, 4; subsidies, 61n53; values of securities, 1; war on terror, 51n10. *See also* Antitrust issues; *individual airlines*

Airline industry, regulation and deregulation: *1978* deregulation, 1, 7, 60n46; effects of, 3, 4, 23, 47–48, 51n10, 60n46; regulatory vs. antitrust conditions, 14–17; success of deregulation, 3–4, 6–7, 18, 52n16

American Airlines, 53n17, 55n21, 57n30, 60n45

Ameritech, 59n44, 66n79

Antitrust issues: airline industry and, 7, 11; alliances, 12–13, 17–19; enforcement, 47; frequent flyer programs and, 14; legislative constraints, 3; mergers, 8, 14–19; regulatory and antitrust conditions, 14–17, 41–42

Appalachian Coals decision (*1933*), 11
Aron, Debra, 35–36, 63n62, 70n86,
    76n104
Association for Local Telephone
    Services, 76n104
AT&T: as a competitor, 35, 38,
    69n85, 75n99; divestiture, 23–24;
    in Illinois, 76n104, 78n105; rates
    and pricing, 37–38, 69n85, 75n99,
    76n104; research support, 22;
    reselling Verizon products, 67n84.
    See also Baby Bells; *individual*
    telephone companies

Baby Bells: AT&T divestiture, 23–2;
    competition and, 39–40, 42,
    64n64, 78n106, 79n110; service
    offerings, 25, 26, 29, 61n54. *See*
    also AT&T
Baker, Jonathan, 57n30, 57n33
Bell South, 40, 75n99
Branson, Richard, 51n11
Breyer, Stephen A., 29, 76n104
"Bribing Customers to Leave and
    Calling It 'Competition'" (Kahn),
    38
British Airways, 53n17, 57n30
Brueckner, Jan K., 10

Cable service, 34, 43, 44, 70n86,
    81n120
California, 35, 79n108
Canada, 61n54, 66n81
Canadian Radio-Television and
    Telecommunications Commission,
    66n81
Carter (Jimmy) administration, 5
Chrysler Corporation, 5
Civil Aeronautics Act of *1938*, 11

Civil Aeronautics Board (CAB), 1,
    6–7, 11
Clayton Act, 15
CLECs. *See* Competitive local
    exchange companies
Colorado Public Service Commission,
    74n98
Competition: in airline alliances, 10,
    12, 13–17, 56n27; code-sharing, 10,
    16–17; economic factors in, 11; of
    low-cost carriers, 6; mergers and
    consolidation, 15, 53n17; pricing
    and charges, 69n85; subsidies and,
    38; superiority of, 2, 78n106; in
    telecommunications, 21–22, 25–26,
    31, 33–35. *See also* Antitrust issues
Competitive local exchange companies
    (CLECs): comparison with wireless,
    80n111; competition and services,
    25–26, 40–41, 67n84, 71n87,
    79n108; complaints to public utility
    commissions, 42; decline and
    defaults, 64n67, 65n70, 71n87;
    definition of, 25; investment by,
    25–26, 32, 63n62; lines and line
    sharing, 43–44, 67n84, 70n86; rates
    and pricing, 71n87, 79n108; use of
    UNE–Ps, 63n62, 71n91, 80n111.
    See also Telecommunications
Computer reservation systems (CRS),
    16
Continental Airlines: conditions of
    alliance with Delta and Northwest,
    15–18, 60n47; effects of alliance
    with Delta and Northwest, 11–12,
    13–14, 56n27, 57n30, 58n39;
    merger with Northwest, 57n33;
    opposition to UAL–US Airways
    merger, 8; purpose of alliance with

JOINT CENTER

AEI-BROOKINGS JOINT CENTER FOR REGULATORY STUDIES

Executive Director
Robert W. Hahn

Director
Robert E. Litan

Fellows
Robert W. Crandall
Christopher C. DeMuth
Judith W. Pendell
Scott J. Wallsten
Clifford M. Winston

In order to promote public understanding of the impact of regulations on consumers, business, and government, the American Enterprise Institute and the Brookings Institution established the AEI-Brookings Joint Center for Regulatory Studies. The Joint Center's primary purpose is to hold lawmakers and regulators more accountable by providing thoughtful, objective analysis of relevant laws and regulations. Over the past three decades, AEI and Brookings have generated an impressive body of research on regulation. The Joint Center builds on this solid foundation, evaluating the economic impact of laws and regulations and offering constructive suggestions for reforms to enhance productivity and welfare. The views expressed in Joint Center publications are those of the authors and do not necessarily reflect the views of the Joint Center.

COUNCIL OF ACADEMIC ADVISERS